COAL WASTE IMPOUNDMENTS

Risks, Responses, and Alternatives

Committee on Coal Waste Impoundments
Committee on Earth Resources
Board on Earth Sciences and Resources
Division on Earth and Life Studies
National Research Council

NATIONAL ACADEMY PRESS
Washington, D.C.

NOTICE: The project that is the subject of this report was approved by the Governing Board of the National Research Council, whose members are drawn from the councils of the National Academy of Sciences, the National Academy of Engineering, and the Institute of Medicine. The members of the committee responsible for the report were chosen for their special competences and with regard for appropriate balance.

This study was supported by an agreement between the National Academy of Sciences and the Mine Safety and Health Administration. Any opinions, findings, conclusions, or recommendations expressed in this publication are those of the author(s) and do not necessarily reflect the view of the organizations or agencies that provided support for this project.

International Standard Book Number 0-309-08251-X

Library of Congress Control Number 2001097318

Additional copies of this report are available from:
National Academy Press
2101 Constitution Ave., NW
Box 285
Washington, DC 20055
800–624–6242
202–334–3313 (in the Washington metropolitan area)
http://www.nap.edu

Cover: Original illustration by Van Nguyen.

Printed in the United States of America

THE NATIONAL ACADEMIES

National Academy of Sciences
National Academy of Engineering
Institute of Medicine
National Research Council

The **National Academy of Sciences** is a private, nonprofit, self-perpetuating society of distinguished scholars engaged in scientific and engineering research, dedicated to the furtherance of science and technology and to their use for the general welfare. Upon the authority of the charter granted to it by the Congress in 1863, the Academy has a mandate that requires it to advise the federal government on scientific and technical matters. Dr. Bruce M. Alberts is president of the National Academy of Sciences.

The **National Academy of Engineering** was established in 1964, under the charter of the National Academy of Sciences, as a parallel organization of outstanding engineers. It is autonomous in its administration and in the selection of its members, sharing with the National Academy of Sciences the responsibility for advising the federal government. The National Academy of Engineering also sponsors engineering programs aimed at meeting national needs, encourages education and research, and recognizes the superior achievements of engineers. Dr. Wm. A. Wulf is president of the National Academy of Engineering.

The **Institute of Medicine** was established in 1970 by the National Academy of Sciences to secure the services of eminent members of appropriate professions in the examination of policy matters pertaining to the health of the public. The Institute acts under the responsibility given to the National Academy of Sciences by its congressional charter to be an adviser to the federal government and, upon its own initiative, to identify issues of medical care, research, and education. Dr. Harvey V. Fineberg is president of the Institute of Medicine.

The **National Research Council** was organized by the National Academy of Sciences in 1916 to associate the broad community of science and technology with the Academy's purposes of furthering knowledge and advising the federal government. Functioning in accordance with general policies determined by the Academy, the Council has become the principal operating agency of both the National Academy of Sciences and the National Academy of Engineering in providing services to the government, the public, and the scientific and engineering communities. The Council is administered jointly by both Academies and the Institute of Medicine. Dr. Bruce M. Alberts and Dr. Wm. A. Wulf are chairman and vice chairman, respectively, of the National Research Council.

www.national-academies.org

COMMITTEE ON COAL WASTE IMPOUNDMENTS

FRANKLIN M.ORR, Jr., *Chair,* Stanford University, California
GARY A.DAVIS, University of Tennessee, Knoxville
BARBARA A.FILAS, Knight Piesold Consulting, Denver, Colorado
C.DAVID HENRY, Beard Technologies, Inc., Pittsburgh, Pennsylvania
NORBERT R.MORGENSTERN, University of Alberta *(emeritus),* Independent Consultant, Edmonton, Alberta
DAVID A.NEWMAN, Appalachian Mining and Engineering, Inc., Lexington, Kentucky
RAJA V.RAMANI, Pennsylvania State University, University Park
ROBERT L.SCHUSTER, U.S. Geological Survey *(emeritus),* Independent Consultant, Denver, Colorado
MADAN M.SINGH, Engineers International, Inc., Scottsdale, Arizona
DON W.STEEPLES, University of Kansas, Lawrence
CLINTON L.STRACHAN, Shepherd Miller Inc., Fort Collins, Colorado
RICHARD J.SWEIGARD, University of Kentucky, Lexington
JACK TISDALE, Mine Safety and Health Administration *(emeritus),* Independent Consultant, Chesapeake, Virginia
DAVID R.WUNSCH, State Geologist of New Hampshire, Concord

NRC Staff

TAMARA L.DICKINSON, Study Director
KRISTEN L.KRAPF, Research Associate
KERI H.MOORE, Research Associate
MONICA R.LIPSCOMB, Research Assistant
KAREN L.IMHOF, Senior Project Assistant
WINFIELD SWANSON, Editorial Consultant

COMMITTEE ON EARTH RESOURCES

SUSAN M.LANDON, *Chair,* Thomasson Partner Associates, Denver, Colorado
JAMES C.COBB, Kentucky Geological Survey, Lexington
VICKI J.COWART, Colorado Geological Survey, Denver
GRAHAM A.DAVIS, Colorado School of Mines, Golden
P.GEOFFREY FEISS, College of William and Mary, Williamsburg, Virginia
MURRAY W.HITZMAN, Colorado School of Mines, Golden
JAMES M.McELFISH, JR., Environmental Law Institute, Washington, DC
DIANNE R.NIELSON, Utah Department of Environmental Quality, Salt Lake City
THOMAS J.O'NEIL, Cleveland-Cliffs, Inc., Cleveland, Ohio
RICHARD J.STEGEMEIER, Unocal Corporation, Anaheim, California
HUGH P.TAYLOR, JR., California Institute of Technology, Pasadena
R.BRUCE TIPPIN, North Carolina State University, Asheville
MILTON H.WARD, Ward Resources, Inc., Tucson, Arizona
LAWRENCE P.WILDING, Texas A&M University, College Station
PHILLIP MICHAEL WRIGHT, Idaho National Engineering and Environmental Laboratory, Idaho Falls

NRC Staff

TAMARA L.DICKINSON, Senior Program Officer
KERI H.MOORE, Research Associate
KAREN L.IMHOF, Senior Project Assistant

BOARD ON EARTH SCIENCES AND RESOURCES

RAYMOND JEANLOZ, *Chair,* University of California, Berkeley
JOHN J.AMORUSO, Amoruso Petroleum Company, Houston, Texas
PAUL B.BARTON, JR., U.S. Geological Survey, *(emeritus),* Reston, Virginia
DAVID L.DILCHER, University of Florida, Gainesville
BARBARA L.DUTROW, Louisiana State University, Baton Rouge
ADAM M.DZIEWONSKI, Harvard University, Cambridge, Massachusetts
WILLIAM L.GRAF, University of South Carolina, Columbia
GEORGE M.HORNBERGER, University of Virginia, Charlottesville
SUSAN KIEFFER, S.W.Kieffer Science Consulting Inc., Bolton, Ontario
DIANNE R.NIELSON, Utah Department of Environmental Quality, Salt Lake City
JONATHAN G.PRICE, University of Nevada, Reno
BILLIE L.TURNER, II,Clark University, Worcester, Massachusetts

NRC Staff

ANTHONY R.DE SOUZA, Director
TAMARA L.DICKINSON, Senior Program Officer
DAVID A.FEARY, Senior Program Officer
ANNE M.LINN, Senior Program Officer
PAUL M.CUTLER, Program Officer
LISA M.VANDEMARK, Program Officer
KRISTEN L.KRAPF, Research Associate
KERI H.MOORE, Research Associate
MONICA R.LIPSCOMB, Research Assistant
JENNIFER T.ESTEP, Administrative Associate
VERNA J.BOWEN, Administrative Assistant
YVONNE P.FORSBERGH, Senior Project Assistant
KAREN L.IMHOF, Senior Project Assistant
SHANNON L.RUDDY, Project Assistant
TERESIA K.WILMORE, Project Assistant
WINFIELD SWANSON, Editorial Consultant

This report has been reviewed by individuals chosen for their diverse perspectives and technical expertise, in accordance with procedures approved by the NRC's Report Review Committee. The purpose of this independent review is to provide candid and critical comments that will assist the institution in making its published report as sound as possible and to ensure that the report meets institutional standards for objectivity, evidence, and responsiveness to the study charge. The review comments and draft manuscript remain confidential to protect the integrity of the deliberative process. We wish to thank the following individuals for their review of this report:

Richard G.Almes, Almes and Associates, Inc.
Joseph Cook, Mine Safety and Health Administration (retired)
Albert W.Deurbrouck, Department of Energy (retired)
Thomas V.Falkie, Berwind Natural Resources Corporation (retired)
J.Steven Gardner, Engineering Consulting Services, Inc.
James M.McElfish, Jr., Environmental Law Institute
James K.Mitchell, Virginia Polytechnic Institute *(emeritus)*
John Morgan, Morgan Worldwide
Gary R.Olhoeft, Colorado School of Mines
Syd S.Peng, West Virginia University
William W.Woessner, University of Montana

Although the individuals listed above have provided many constructive comments and suggestions, they were not asked to endorse the conclusions and recommendations nor did they see the final draft of the report before its release. The review of this report was overseen by George M.Hornberger, University of Virginia, Ernest H.Ern Professor of Environmental Sciences, and Paul B. Barton, Jr., United States Geological Survey, Geologist Emeritus. Appointed by the National Research Council, they were responsible for making certain that an independent examination of this report was carried out in accordance with institutional procedures and that all review comments were carefully considered. Responsibility for the final content of this report rests entirely with the authoring committee and the institution.

Contents

Executive Summary

On October 11, 2000, near Inez, Kentucky, a breakthrough occurred in which a 72-acre surface impoundment of waste materials of the Martin County Coal Corporation released approximately 250 million gallons of slurry into a nearby underground coal mine. The slurry flowed through the mine and into nearby creeks and rivers, flooding stream banks to a depth of 5 feet. Although the spill caused no loss of human life, environmental damage was significant, and local water supplies, taken from the rivers, were disrupted for days. This incident caused Congress to request the National Research Council to examine ways to reduce the potential for similar accidents in the future. To conduct this study, the National Research Council appointed the Committee on Coal Waste Impoundments. The committee held a total of 10 meetings between March and July 2001, eight of which included a town meeting to gain input from citizens of local communities.

The charge to the committee includes three major components. First, the committee was to examine engineering practices and standards currently being applied to coal waste impoundments and to consider options for evaluating, improving, and monitoring the barriers that retain coal waste impoundments. Second, the committee was charged with evaluating the accuracy of mine maps and exploring ways to improve surveying and mapping of underground mines to delineate more accurately how underground mines relate to current or planned slurry impoundments. The third task was to evaluate alternative technologies that could reduce the amount of coal waste generated or allow productive use of the waste. The committee also examined alternative disposal options for coal slurry.

It is important to recognize that this charge specifically directs the committee to focus its analysis on the engineering and characterization of coal waste impoundments. The committee was not asked to consider other factors that are related to potential impacts of disposing of coal waste in an impoundment, or any other disposal option. For example, these factors might include potential long-term effects on water quality; land-use issues, including long-term stewardship of closed impoundments; and economic and cost-benefit analyses of alternatives. The committee also was not asked to evaluate the risks of individual impoundments, examine the qualifications and training of inspectors, or comment on coal mining policy issues not directly related to impoundments. Although important, such issues are well beyond the charge to this committee. Furthermore, a comprehensive analysis of these issues would require considerably more time than was available for the present study.

Advances in mining technology have increased productivity at the expense of quality of the run-of-the-mine product, both in terms of included impurities and a greater proportion of fine-grained coal. This is particularly true for the coal from eastern coal fields, and upgrading of the product to meet requirements of power plants and other users is now a common practice. The treatment of the raw coal can produce as much as 50 percent waste as coarse included rock and as fine-grained coal and mineral matter. Coal waste slurry is one of the refuse streams and is composed mainly of fine coal, small particles of rock, and clay suspended in water. Coal waste slurry is usually disposed of by pumping it into an impoundment, where particles are allowed to settle. Most impoundments in Appalachia utilize the natural topography to form the storage basin that will contain the slurry. This is often accomplished by constructing an embankment in a valley or watershed to complete the basin structure used for storage. Impoundments are often located in steeply sloping valleys.

Coal waste slurry facilities have been involved in several accidents since the 1972 Buffalo Creek incident, where a coal waste impounding structure collapsed, killing 125 people, injuring 1,100, and leaving more than 4,000 homeless. The majority of the incidents involve failure in the basin area. Inaccurate mine maps and inadequate characterization of the basin area most likely contributed to at least some of these incidents. This report, in an effort to identify potential causes of failure, evaluates current practice in site characterization, basin placement, and embankment construction, maintenance, and monitoring. It concludes with a discussion of alternative methods of coal use and waste disposal. Technical terms used in the text are defined in the glossary (Appendix C).

ENGINEERING STANDARDS, BARRIER STABILITY AND MONITORING

The Mining Enforcement and Safety Administration developed standards for impoundments and refuse piles after the 1972 Buffalo Creek incident. One of the purposes of the Federal Mine Safety and Health Act of 1977 was to establish mandatory standards to protect the health and safety of miners. This charge is important in assessing the Mine Safety and Health Administration's (MSHA's) regulatory responsibility and authority associated with potential impoundment breakthroughs and failures that do not directly affect the health and safety of miners but may affect persons, communities, or ecosystems downstream. On the other hand, the Office of Surface Mining's (OSM's) legislative and regulatory direction under the Surface Mining Control and Reclamation Act of 1977 (SMRCA) to protect society and the environment from the adverse effects of surface coal mining operations, complements the intent of the Federal Mine Safety and Health Act.

The committee examined the regulations by MSHA and OSM and its state delegate programs that directly relate to the design, construction, operation, and closure of refuse impoundments, as well as to alternative refuse disposal techniques. The committee also examined other federal statutes that may relate to refuse impoundments or alternatives (e.g., the Clean Water Act). The regulatory structure that covers many aspects of the design, construction, and operation of coal waste impoundments is extensive. The MSHA and OSM regulatory language addresses in detail the engineering and stability aspects of the embankment; however, there is little reference either to the basin area or to requirements of the engineering design for other areas of the facility. The impoundment operator and professional engineer are responsible for providing information about underground mine workings, including the depth and extent of the workings. However, no regulation or standard industry practice instructs or guides the engineer in this task, and there is no procedure for an independent verification of the information submitted. In addition, there is no regulation requiring an evaluation of the breakthrough potential of impoundments. However, in practice, MSHA and OSM appear to have the jurisdiction to require an evaluation of breakthrough potential through indirect regulatory language. The committee concludes that while the regulatory review of a proposed impoundment is detailed with respect to the embankment, the regulatory review of the impoundment basin has been less rigorous. The authority for review of the basin characterization and design appears to be covered only in general language authorizing investigation of all relevant issues with respect to the impoundment. **The committee recommends that MSHA and OSM should have clear authority to review basin design.** It is not evident to the committee whether specific legislation to authorize more detailed examination of basin issues is required, or whether these issues can be handled by additional rulemaking under existing authority.

The committee examined in particular the elements of design, operation, and reclamation of coal refuse impoundment systems in the Appalachian coal region. The principles that govern the design of structures, which were promulgated in response to the Buffalo Creek disaster, are well understood and fully documented. An embankment built to contain coal mine waste is similar to an embankment built to contain mine waste from other extractive industries. Reviews of failure modes in other extractive industries are relevant to identifying potential failure modes of coal waste impoundments. Given the fact that some modern dams in other extractive industries have failed, the committee concludes that it is essential that MSHA and OSM stay current by ensuring that design criteria reflect the latest experience from all segments of the mining industry. Although the committee has not identified any deficiencies, it is a matter of due diligence that MSHA and OSM and industry employ the best available current technology. **The committee recommends that MSHA and OSM continue to adopt and promote the best available technology and practices with regard to the site evaluation, design, construction, and operation of impoundments.** For example, MSHA and OSM should commission periodic reviews of existing technical procedures and practices, with particular attention to the basin. Results of the reviews should be disseminated to industry. Based on the outcome, MSHA and OSM may need to revise guidelines to establish minimum expectations and levels of investigation for site characterization, design, construction, operation, and closure of coal refuse impoundments.

Embankments can fail in a variety of ways, including slope instability, liquefaction, and foundation failure. Seepage through embankments can lead to failure by internal erosion. Overtopping of an embankment can cause substantial erosion of the crest, which, if left uncontrolled, will work progressively downward, releasing water and coal refuse downstream. While continued vigilance concerning design, construction, and operation of embankments is clearly warranted, the committee concludes that the largest uncertainties remain in the characterization of the basin area and, therefore, in the mitigation of risks associated with the breakthrough potential. The potential for underground coal mine workings to be near an impoundment is a factor in the design of new and in modifications to existing coal waste impoundments in Appalachia. The relative elevations of local drainage and slurry height, with respect to underground mines, can be a critical design issue. The stream channel at the base of the impoundment basin defines the approximate level of local drainage. Coal seams and mines that do not crop out above the level of the stream channel are termed below-drainage; whereas, those that crop out along the valley wall above the stream channel are termed above-drainage. Existing impoundments with above-drainage mine workings, where the slurry elevation does not exceed the level of the coal mine workings, can relatively easily incorporate mitigative measures for these workings in their design. Above-drainage coal mine workings in existing impoundments, where the slurry elevation exceeds the level of the mine workings, are the most challenging in the design and operation of a facility.

If the outcrop coal barrier is of insufficient width, or the overburden above the mined area is too thin, impounded water can break into a mine. Conversely, an inundated mine under high hydraulic head can introduce a large volume of water into the impoundment. Should a blow-out occur elsewhere in the watershed, above the pool level, an inflow of substantial amounts of water into the impoundment can result. This, in turn, may result in overtopping of the embankment or damage to the principal and emergency spillways. Synergistic reactions among geologic and hydrogeologic conditions can compound the instabilities created by any one of the failure modes. Currently, no federal regulations address the width of outcrop barrier that should be left during underground coal mining. OSM has studied the problem of outcrop barriers but has released no conclusions to date. **The committee recommends that MSHA and OSM jointly pursue the issue of outcrop coal barrier width and overburden thickness and its competence and develop minimum standards for them.**

If slurry from an impoundment leaks into active or abandoned mine workings or may do so, bulkheads or seals may be constructed to preclude the water from escaping into the outside environment. Many mitigative measures can be designed using established procedures; however, bulkheads designed to support high hydrostatic pressure present a different kind of problem. **The committee recommends that MSHA review its current practice and develop guidelines for the design of bulkheads intended to withstand hydraulic heads associated with slurry impoundments.** The bulkhead should be constructed from material that can withstand water action without deterioration in the presence of the various chemicals in the impoundment water. Furthermore, the bulkhead should be suitably anchored in competent, unfractured strata. If such an area is not available, pressure grouting may be needed. Deterioration of the anchoring strata can be a major structural problem where the bulkhead is keyed into water-sensitive, clay-bearing strata. The size, integrity, and strength of the surrounding coal pillars, roof, and floor are critical to successful sealing. Generally, seals constructed for ventilation cannot withstand water pressure.

The committee concludes that selecting the appropriate mitigative measures relies strongly on reliable basin characterization. **The committee recommends that MSHA and OSM develop and promulgate guidelines**

for the site evaluation, design, construction, and operation of basins. They should be comparable in scope to the guidelines used in embankment design.

Monitoring is a critical component in the construction and operation of coal slurry impoundments. Measurements are commonly made to detect surface displacements, internal movement, pore pressure, groundwater level, surface water discharge, and subsurface movement. Procedures in place for monitoring the embankment, which include visual inspection and instrumentation, appear to be performing as envisioned in the regulations MSHA implemented. For monitoring to be successful, it should be applied to all potential failure modes. The committee believes that there are opportunities for additional continuous monitoring that may offer timely warning in case of impending failure of an embankment or basin. **The committee recommends that MSHA and OSM consider requiring additional continuous monitoring in specific instances and evaluate automation of monitoring instruments.**

SITE CHARACTERIZATION

Key to assessing the potential for breakthrough of coal slurry into underground mine workings is knowing the extent of those workings with respect to the ground surface in the impoundment basin area. The committee examined several aspects of the topic of site characterization, including geology, hydrogeology, the accuracy of surface and mine maps, and methods for delineating the extent of underground mine workings in situations where maps are nonexistent or inaccurate. A particular problem is old surveys and maps, some of which have been lost or destroyed. If unknown mine workings are present, the impoundment could suffer unexpected structural failure. In areas where impoundments are constructed near known or suspected underground mines, vertical and horizontal barrier distances between the mined area and the impoundment may not be accurate. Current regulations require closed-loop mine surveys, but surveys for many older mines were not closed. Furthermore, underground mine surveys may have been based on a foreman's notes or sketches that lack a reference point or a recognized coordinate system that would allow accurate location. These shortcomings are more common in small mines or room-and-pillar mines where a number of short panels were driven and extracted. **Therefore, the committee recommends that MSHA work with OSM and state agencies to establish standards for mine surveying and mapping. These should include the following:**

- **Determining surface coal outcrop locations by aerial topographic measurements, where adjacent to existing or proposed refuse impoundments,**
- **Implementing a coordinated and assertive approach to collecting and archiving mine maps,**
- **Scanning paper copies of mine maps into electronic data files upon receipt,**
- **Setting standards for minimum closure error for all underground closed-loop surveys and that a closed-loop survey be maintained within a standard distance (to be determined by MSHA),**
- **Recording the depth of the last cut taken to a level of accuracy to be determined by MSHA,**
- **Using state plane coordinates or latitude and longitude, and bottom-of-seam elevations as the map base reference,**
- **Listing appropriate coordinate transformation equation(s) on the mine map,**
- **Adding a qualifying statement to accompany any coordinate transformation that is based upon the alignment of surface features,**
- **Improving and maintaining the location of surface controls,**
- **Determining which mine permit documents should be retained, in what form, and for how long,**
- **Avoiding the use of coal seam names as the sole basis for determining the vertical location of an abandoned mine.**

When no mine maps can be found or there is reason to doubt their accuracy, additional investigation to locate underground mine workings is warranted. This can be expensive and time consuming. Because of the time and expense and remaining inherent uncertainty associated with extensive drilling, remote sensing and geophysical methods have been employed to search for abandoned coal mines. The objective of geophysical surveys is to determine the physical characteristics of a three-dimensional volume of earth material, including the presence of voids. Since no geophysical technique is capable of performing optimally under all geologic and topographic conditions, multiple geophysical techniques may be necessary to reduce the probability for error to an acceptable level. Drilling is still necessary to confirm and calibrate interpretations of geophysical and remote sensing data. The absence of evidence of a mine is not evidence of absence of a mine, and there are many opportunities for error in the modeling and geophysical surveys needed to detect voids. The committee concludes that geophysical techniques can be useful in coal mine void detection, especially the use of seismic surface waves, seismic reflection, ground-penetrating radar, and electrical resistivity methods. The committee also concludes that geophysical techniques have been underutilized in the coal-mining industry and could benefit from additional research. **The committee recommends that demonstration projects using modern geophysical techniques be funded, and that the results be widely conveyed to the mining industry and to government regulatory personnel through workshops and continuing education.** Continuing education could include the opportunity to attend short, courses and seminars that present the latest technology along with case histories to support its use.

ALTERNATIVE TECHNOLOGIES

Coal waste impoundments are one of the waste disposal options of the present system of mining and preparing coal for energy production. To assess thoroughly other alternatives, the entire system of mining, preparation, refuse disposal, transportation, and power generation should be explored through an in-depth life-cycle assessment, including cost assessment, with the goal of optimizing the system to generate less fine coal waste while maintaining the performance and economics of the system. However, benefits may be difficult to realize because of differences in interests and perceptions between the mining industry and the utility industry, and because of the resistance to change embedded in these mature industries. **The committee recommends that the total system of mining, preparation, transportation, and utilization of coal and the associated environmental and economic issues be studied in a comprehensive manner to identify the appropriate technologies for each component that will eliminate or reduce the need for slurry impoundments while optimizing the performance objectives of the system.** The committee concludes that a similar analysis of the waste use and disposal technologies that make up the coal system would have value. **The committee recommends incorporating life-cycle assessment of the costs and environmental impacts of the alternatives to evaluate them on a more objective, comprehensive basis. In addition, a detailed analysis of the economic and environmental impact of the various policy alternatives should be performed.**

The opportunities for reducing slurry volume include mining alternatives and coal processing alternatives. However, modern methods of surface and underground coal mining offer a limited possibility for quality control during mining. Slurry volume can be reduced by improving fine coal recovery, minimizing the mass of solids for disposal, and dewatering. Many dewatering technologies are currently available for specific applications, though none is likely to be universally applicable. The committee believes that equipment vendors' current research and development will lead to improvements in these technologies and that operators of coal waste impoundments should monitor them carefully.

Slurry refuse can be used directly for power generation, either in conventional boilers or in advanced combustion and gasification technologies. Some of these technologies can reduce cleaning requirements for coal. However, the use of low quality coal feed will increase the amount of waste generated at the power plant. The committee concludes that technologies for utilization of fine coal waste for electricity generation in conventional coal-fired power plants are available. These technologies offer near-term opportunities for the reduction of fine coal waste disposed of in impoundments. However, the coal produced is more expensive than cleaned coal, as a result of capital and operating costs of additional equipment, and, in the case of coal water slurry, the additional cost of transportation. To compare technologies fully, the avoided costs of slurry impoundments must be included in the analysis.

Fluidized-bed combustion and gasification show promise for recovering the heat content of fine coal waste while avoiding some of the operational problems that limit use of coal fines in conventional pulverized coal-fired boilers. The committee concludes that the burning of fine coal waste in advanced combustion technologies, such as fluidized-bed combustion and gasification, is an alternative that shows considerable long-term promise. Atmospheric fluidized-bed units are already in use for combustion of fine coal waste slurries from both preparation plants and old slurry impoundments, but they have not gained wide usage. Pressurized fluidized-bed technologies offer improved efficiency over atmospheric technologies but have not been utilized in full-scale applications for burning fine coal waste. Gasification technologies are also promising for coal water slurries, because they operate more efficiently, and because emerging technologies can utilize the water from the slurry as a steam source required for the gasifier. Further research is needed on the use of fine coal waste slurries as feeds, and incentives may be useful if these technologies are to be widely incorporated for fine coal waste combustion. Even though coal combustion wastes from power plants are already being used for a number of purposes, safe handling of coal combustion waste from these advanced combustion technologies should be studied further.

Methods are available for the disposal of coal slurry other than in impoundments, including both surface and underground options. Alternative surface methods include incised ponds, slurry cells, combined refuse piles, and co-disposal of fine and coarse refuse. In many instances, these methods are influenced by topography, geology, and mining and coal preparation characteristics, and, therefore, their application is site specific.

Incised ponds, which are similar to slurry impoundments but without an embankment, are designed to accept any form of coal waste including slurry. This method of surface disposal is most common in the Midwest where area surface mining is practiced on flat topography. Area surface mining produces long end-cuts and inclines, which are usually allowed to fill with water to form a permanent lake. These excavations are at or below the level of surface drainage and do not present a danger of sudden failure if used for slurry disposal.

Slurry cells are designed to impound less than 20 acre-feet of slurry. The advantage of this method over conventional cross-valley impoundments is that each cell is small and self-contained and can be designed according to the strength properties of the coarse refuse. The main disadvantage in steep terrain is the limited availability of flat land to construct the cells. Another disadvantage is that slurry cell operations are not compatible with a high coal preparation production rate.

Combined refuse piles consist of fine refuse from a static thickener that has been mechanically dewatered and combined with coarse refuse. Mechanical dewatering to yield the required moisture level is frequently a problem. In addition, the dewatered material is difficult to handle and compact. Combined refuse is expensive because of the high cost of mechanical dewatering and the potential need for chemical additives to stabilize the combined refuse. This method is best suited to flat land.

Co-disposal involves the combination of fine refuse from the static thickener with coarse refuse. This method requires less total storage volume than separate fine and coarse disposal methods, and the refuse stabilizes more quickly than typical slurry. Co-disposal has been used primarily in sparsely populated areas with low annual rainfall. Questions remain about its suitability for steep hills with high annual rainfall.

If an effective dewatering approach, such as paste thickening, is used, the resulting waste can be disposed of by thickened high-density residue stacking. Although used in other extractive industries, this process has seen limited use in the coal industry for the disposal of fine coal refuse. Three considerations—land availability, steep terrain, and cost—hamper applying unsupported thickened high-density residue stacking to fine coal refuse disposal. This method is best suited to areas where the slope of the land is less than 5 percent. The lower throughput rate of deep cone thickeners compared with that of standard thickeners may significantly affect the economic feasibility of the method.

For the underground options, two primary methods for injecting fine coal refuse into underground mines are controlled flushing, where the underground workings are accessible, and blind or uncontrolled flushing, where the underground workings have been abandoned or have caved in. However, several issues are both important and independent of the method of slurry injection. For example, it is essential to have an adequate supply of water. This is especially true when water is not being recaptured from the underground workings. Additional issues include surface ownership, permits, surface layout, and surface drainage. Filling above-drainage mine workings with slurry may increase hydraulic head on the coal barriers and result in a blow-out, making evaluation of mine workings above a surrounding stream valley critical. The accuracy of mine maps must be ascertained, and the underground barriers must be assessed for adequacy to contain the slurry. Mines below the surrounding natural drainage level offer more secure underground disposal sites.

The committee concludes that although there are alternatives to disposing of coal waste in impoundments, no specific alternative can be recommended in all cases. Acceptable alternatives are highly dependent upon regional and site-specific conditions. Also, the alternatives that have been identified are in varying stages of technological development and implementation. One of the factors limiting implementation to this point has been the costs associated with the various alternatives. Additional research is needed to develop these alternatives further and to evaluate the economics of these processes. **The committee recommends that a screening study be conducted that (1) establishes ranges of costs applicable to alternative disposal options, (2) identifies best candidates for demonstration of alternative technologies for coal waste impoundments, and (3) identifies specific technologies for which research is warranted.** Input from MSHA and OSM regarding regulatory issues will be valuable to such a study. **The committee recommends that the use of economic incentives be explored as a way of encouraging the development and implementation of alternatives to slurry impoundments.** The development of incentives should be based on the full range of the portfolio of technologies as well as the economics of the technologies. The incentives should be linked directly to the reduction in slurry production or the utilization of slurry.

One method for reducing the volume of material in older slurry impoundments is to recover or remine the fine coal. Older impoundments contain significant amounts of coal refuse with recoverable energy value. However, as processing technologies and the capacity of dewatering equipment have improved, the proportion of coal with high energy value being deposited in slurry impoundments has decreased. In many cases, the finer slurry materials being disposed of today with less recoverable and marketable coal are placed over the more amenable, profitable, and recoverable slurry. Typically, if an impoundment contains at least 1 million tons of in-situ slurry, a recovery rate of at least 30 percent of a marketable fine coal product (300,000 tons) from the slurry could prove to be a profitable venture. The -committee concludes that as advances are made in the use of low value coal or coal water slurry, remining of slurry impoundments can be an attractive source for fuel supply.

ADDITIONAL RECOMMENDATIONS

In its deliberations, the committee identified several issues that cut across elements of the statement of task and some related issues that warrant additional study.

MSHA currently uses two systems to classify coal waste impoundments. One system classifies impoundments as high, medium, or low hazard, based upon the magnitude of the potential consequences of failure of the embankment structure. If communities and structures are immediately downstream, the embankment would be classified as high hazard, regardless of its likelihood to fail. A second system addresses the potential for the unintentional release of water or slurry from impoundments into active or abandoned mines. These ranking systems are based on the proximity of the basin to underground workings as well as the potential downstream impacts if a basin were to fail. The second classification comes closer to the standard definition of risk as the product of hazard (the likelihood of failure) and consequences (such as loss of life, environmental damage). The committee concludes that using different hazard classification methodologies for embankments and basins is inappropriate. **Therefore, the committee recommends that: (1) MSHA and OSM review activities related to risk assessment for existing impoundments (including both embankments and basins) to ensure that they are consistent and that they distinguish appropriately between hazard and consequence assessment in the methodologies adopted; and (2) MSHA and OSM establish a single, consistent system, which should be used to assign both embankments and basins to risk categories.** The ranking should be based on the appropriate combination of hazards and consequences. The committee believes that this can be accomplished using qualitative risk assessment techniques.

A consistent risk assessment system would allow decisions to be based on their relative risks. **The committee recommends that MSHA and OSM oversee a thorough assessment of potential mitigation measures for those impoundments that fall in the highest risk category and should determine which mitigation measures should be applied to reduce this risk to an acceptable level.**

The committee also concludes that the design process for impoundments would be improved by a more formal risk analysis. Proposed new impoundments should also be assigned to risk categories, based on a combination of hazards and consequences, as was suggested for existing impoundments. **To maximize the potential for risk reduction, the committee recommends that all impoundment designs be accompanied by a risk analysis utilizing qualitative methods.** Examples of such methods include Potential Problem Analysis and Failure Modes and Effects Analysis.

The committee believes there is a limit to risk tolerance, for both existing and new impoundments. When risk is high, and when mitigation, either through more reliable characterization or barrier construction is impossible, of limited precedent, or so expensive that it is infeasible, then a substantial change in operation of the impoundment is warranted. This may range from minimizing slurry fluidity to ceasing operations. If an impoundment fails risk-assessment criteria and if risk cannot be mitigated it should be phased out or alternatives considered.

In collecting information concerning the design of impoundments and the process by which the design is reviewed by regulatory authorities, the committee heard reports that the review process is undesirably lengthy, commonly exceeding 2 years. In addition, it appears that review times have lengthened considerably in recent years. The committee concludes that timely review is an essential component of an effective regulatory process. An efficient and coordinated regulatory review process can be of substantial benefit to both the applicant and the jurisdictional agencies. A well-coordinated technical review process can ensure that the health and safety of both the miners and the public, and the protection of the environment are ensured in a sensible and streamlined way. **Therefore, the committee recommends that the review process for both new permits and existing permits be overhauled to include the following elements:**

- **A formal joint process that would coordinate the currently fragmented and inefficient collection of reviews into a single process.**

- **Sufficient staff for engineering and other reviews in the agencies that participate in the joint process so that the time required to complete the review can be reduced significantly.**

In its review of information about the impacts of uncontrolled releases of water and slurry from impoundments, the committee found that only very limited information was available concerning the quantities of trace elements present in the slurry and associated water. A common theme at the town meetings was the concern about the quality of ground and surface water. While a detailed review of the environmental impacts of coal waste impoundments was beyond the scope of this study, the committee identified this area as one needing further study. As **a result, the committee recommends that research be performed to identify the chemical constituents contained in the liquid and solid fractions of coal waste, and to characterize the hydrogeologic conditions around impoundments.**

Public concern regarding emergency response and evacuation plans was another recurring theme in public comments made to the committee. Some residents were unaware of emergency evacuation plans; others had seen evacuation plans but disagreed with the logic behind the evacuation routes and would not have used the plan in the event of an emergency. Conversely, coal industry and regulatory agency representatives stated that these plans are being developed and shared with the public through the various community contacts (e.g., local fire departments, police, health care providers). The lack of realistic communication constitutes a fundamental barrier to the industry's ability to make stakeholders aware of the risk associated with coal refuse impoundment construction, operation, and closure and of steps taken to mitigate that risk. The committee concludes that communication concerning coal refuse impoundment risk and emergency response between the industry and the local communities could be improved substantially. The committee suggests that the industry take steps through the appropriate emergency response agencies to address these problems.

SUMMARY

The conclusions and recommendations offered above reflect the committee's judgments concerning ways to improve the design process for coal waste impoundments, ways to improve mapping of mines and the characterization of sites of existing and future impoundments, ways to improve the assessment and mitigation of risks associated with impoundments, and ways to assess options for coal waste slurry. The committee believes that implementation of the recommendations will substantially reduce the potential for uncontrolled release of coal slurry from impoundments, particularly through the mechanism of breakthrough into nearby underground mine workings. In addition, the committee believes that viewing the designs of embankment and basins as well as the entire process of handling and burning coal as systems of interlinked components that operate together is an appropriate way to balance alternatives for creating, handling, and disposing of wastes and to understand and mitigate the impacts of failure of any element in these systems. The safe operation of these systems is a shared responsibility of government and industry that depends on effective engineering design, construction, and operation in addition to appropriate monitoring. With the recommended improvements in each of these areas, the potential for incidents like that in October 2000 at Inez, Kentucky, can be reduced.

1

Introduction

On October 11, 2000, near Inez, Kentucky, a breakthrough occurred in which a 72-acre surface impoundment of waste materials from coal processing at Martin County Coal Corporation released approximately 250 million gallons of slurry into a nearby underground coal mine. The slurry flowed through the mine, into nearby creeks and rivers, flooding stream banks to a depth of 5 feet. The spill caused no loss of human life. However, environmental damage was significant, and local water supplies, taken from the rivers, were disrupted for days. This report develops numerous suggestions and recommendations to reduce the potential for similar accidents in the future. In this chapter, we review the processes that lead to the generation of coal waste and the impoundments used to store it, and then we turn to a description of accidents and incidents involving coal waste facilities. Finally, we review the tasks and activities of this committee.

COAL PRODUCTION AND USE IN THE UNITED STATES

Coal is the largest single source of fuel for domestic energy production. In the United States, 90 percent of the coal produced is used in power plants (Freme and Hong, 2000). Coal accounts for about 33 percent of the total energy production (Chircop, 1999). In 2000, coal accounted for 51.4 percent of electric power generation (Freme, 2001). Industries and manufacturing plants also use coal directly, especially those that produce chemicals, cement, paper, ceramics, and various metal products. On average, about 20 pounds of coal are utilized per day per capita in the United States (Chircop, 1999).

The United States has approximately 26 percent of the world's coal reserves (BP Global, 2001) (Table 1.1). More than 400 coalfields and small deposits underlie a total of 458,600 square miles in 38 states, nearly evenly

TABLE 1.1 Major World Coal Reserves, 2000

Country	Reserves in million short tons	Percentage of world total
United States	246,643	25.1
Former USSR	230,178	23.4
China	114,500	11.6
India	74,733	7.6
South Africa	55,333	5.6
Australia	90,400	9.2
Germany	67,000	6.8
Poland	14,309	1.5
Canada	8,623	0.9
Indonesia	5,220	0.5
United Kingdom	1,500	0.2
Mexico	1,211	0.1
Others	74,561	7.6

SOURCE: Data extracted from BP Global, 2001 .

split between the Eastern and Western regions (Chircop, 1999). Although some 300 different coal beds are mined each year, almost 47 percent of total production comes from just 10 of the largest deposits.

Important coal deposits east of the Mississippi, are found in 10 states (Alabama, Illinois, Indiana, Kentucky, Maryland, Ohio, Pennsylvania, Tennessee, Virginia, and West Virginia). In the West, the Wyodak coal bed, the nation's leading source of coal, underlies part of the Powder River Basin of Wyoming and Montana. Other active Western coal reserves are found in Colorado, Utah, New Mexico, Arizona, and Alaska (Chircop, 1999). In the interior states, coal occurs in several separate basins from Michigan to Texas. According to data from the U.S. Energy Information Administration, coal beds throughout the United States produced 1.08 billion tons of coal in 2000, the 12th consecutive year in which 900 million tons or more were mined. In recent years, U.S. coal production has exhibited the following trends:

- Western mines account for an increasing share of total production (Figure 1.1).
- Fewer coal mines are operating, but those mines are larger.
- Surface mines produce an increased proportion of coal overall.
- Longwall mining produces an increased fraction of coal mined underground.

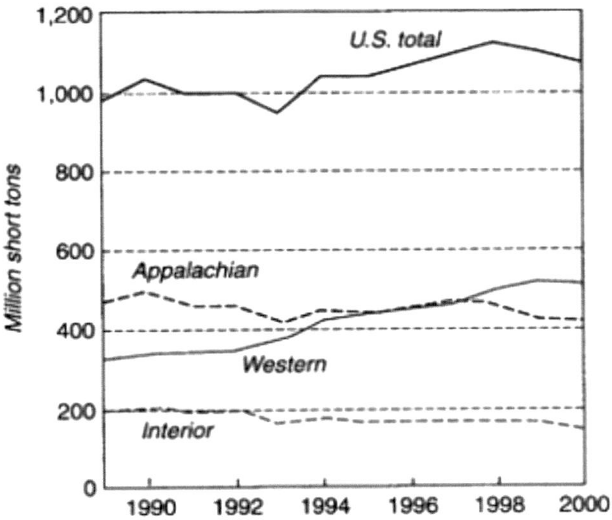

FIGURE 1.1 Coal production by region, 1989–2000. From Energy Information Administration, 2001.

ADVANCED COAL CLEANING

Increased mechanization in the underground coal mining industry has decreased selectivity and increased the volume of refuse. Equipment, such as continuous miners or longwall shearers, often takes roof and floor rock in addition to in-seam partings (Sidebar 1.1). Equipment currently used to mine and transport coal produces more fine coal particles than did earlier equipment. Rotating cutter heads, feeder-breakers, and transfer points in

SIDEBAR 1.1 UNDERGROUND COAL MINING METHODS IN THE UNITED STATES

About 39 percent of coal produced in the United States comes from underground mines (Chircop, 1999). Underground coal is mined by the following methods: conventional, longwall, continuous, and shortwall. Longwall and continuous mining are used effectively in combination (Chircop, 1999).

Conventional mining processes include drilling and blasting the coal. This method, one of the oldest mining methods, can be effective in certain geologic areas. Today, it is used for only 5 percent of underground coal mining (Chircop, 1999).

More than 45 percent of underground coal is mined by longwall mining (Chircop, 1999). This method is gaining in popularity because it can improve coal recovery to 80 percent, and it enhances miners' safety (Peterson et al., 2001). Production rates depend on the width of the block, the thickness of the coal seam, and the technology used to transport the raw coal out of the mine. Rotating drums, steel plows, or mounted shearers traverse back and forth across the block width and excavate blocks 600 to 1,200 feet wide and 5,000 to 7,500 feet long (Chircop, 1999; Peterson et al., 2001). Longwall length capabilities have been steadily increasing: some Western operations now achieve lengths of over 10,000 feet (e.g., Twentymile in Colorado and SUFCO in Utah). The miners and mining equipment are protected by moving hydraulic roof supports called shields (Chircop, 1999). After an area has been mined, the roof collapses.

Continuous mining is a mechanized method utilizing mechanical cutting machinery. Although longwall mining has moved to the forefront (Chircop, 1999; NRC, 2001), continuous mining is still important in coal production. Continuous mining equipment is used to develop the areas for longwall mining. The continuous mining method uses a room-and-pillar system, whereby mined-out "rooms" are supported by coal "pillars." An operator, who maintains visual contact with the machine, can control this machinery remotely, thereby increasing miners' safety.

Shortwall mining refers to mining with a continuous mining machine, moving roof supports, and excavating blocks 150 to 200 feet wide and more than one half mile long (Chircop, 1999). This method is currently not used to any appreciable extent.

conveyor belt systems all break up the coal, producing quantities of fine coal particles whose recovery requires special processing and cleaning techniques.

The characteristics of the in-situ coal (e.g., ash and sulfur content) vary spatially. The product that emerges from a mine often includes, not only randomly distributed impurities and the non-coal material known as partings (see Sidebar 1.2 for a description of the processes that deposit coal and create the impurities), but also material from the roof and floor layers. Thus, the characteristics of mined coal vary considerably.

End-use plants are engineered for optimal combustion to burn feedstock of a particular ash, sulfur, and energy content—requirements rarely met by the run-of-mine coal or by coal from a single source (Figure 1.2). These combustion requirements have been the impetus for all upstream process changes, including the search for coal of low sulfur content, improved coarse and fine coal cleaning processes, in addition to several disposal and environmental laws.

SIDEBAR 1.2 COAL GEOLOGY

Coal is a combustible material consisting of organic matter and minor amounts of inorganic materials. It is derived from a heterogeneous mixture of plant remains and associated minerals, which have undergone chemical and physical changes by geological and biological processes without free access to air. Coal has a highly variable composition, affecting both its chemical and physical properties. Except for the anthracite region in eastern Pennsylvania, coal beds usually occur as nearly horizontal or gently folded strata. A coal seam is a composite of several layers, each of which may consist of a different mixture of coal material and mineral matter. Occasionally, these layers may be completely inorganic, such as shale, or high in mineral matter. Such layers are referred to respectively as partings and bony coal. Coal and associated rocks may contain significant amounts of sulfur, arsenic, and other materials whose presence in the waste engenders environmental concerns.

The depositional environments that produced the coalfields of the Eastern United States were predominantly coastal-deltaic. Large volumes of sediment were deposited from the Appalachian Mountains into rivers, which emptied into bays and coastal seaways. These conditions engendered the development of vast, laterally extensive peat swamps along the coastlines and delta platforms. Periodically, sea level rose and shallow marine environments flooded the swamps, depositing marine shales and limestones. The cyclic repetition of these sedimentary environments has resulted in some of the most complex stratigraphic sequences in the geologic record.

SOURCE: Rice et al., 1979.

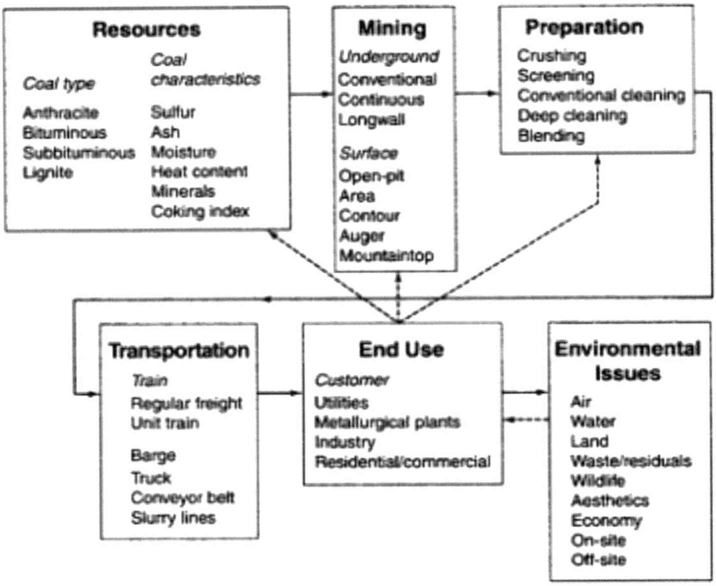

FIGURE 1.2 Coal system components. Dotted arrows indicate important feedback to mining feasibility. Modified from Office of Technology Assessment, 1979.

Limitations on sulfur dioxide (SO_2) emissions from coal-fired power plants have also contributed to the need for advanced coal-cleaning technology. Power plants require coal of consistent quality (e.g., sulfur, ash, and heat content) to comply with these regulations. In an effort to produce coals that allow power plant operators to comply with standards established by the Clean Air Act, various methods of removing pyrite (FeS_2) from the coal have been developed. In the past, much of this material would have entered the combustion chamber with the coal and would have resulted in additional ash. Now, the pyrite is removed, but it adds to the waste the preparation plant generates.

Finally, the quality of coal being mined in the Eastern United States has declined as higher quality reserves have been depleted. Therefore, techniques have been implemented to upgrade the coal product quality.

Previously, coal was cleaned by dry methods; however, a combination of factors, such as particle size, dust, transport, health, safety, and noise, and the better performance of wet processes have contributed to the near abandonment of dry coal cleaning processes in recent years. The increased use of water to control dust in underground mines, and the increased efficiency of wet cleaning methods have continued the sharp decline in dry coal cleaning methods.

Coal preparation (colloquially referred to as "washing") separates noncombustible materials from coal. Processing the run-of-mine coal may include: removing extraneous materials, crushing, sizing, blending coal from several locations, and concentration. A coal preparation plant separates the material it receives into a product stream and a reject stream, which may be further divided into coarse and fine refuse streams. Depending on the source, 20 to 50 percent of the material delivered to a coal preparation plant may be rejected (Leonard, 1991). One of the reject streams is slurry, a blend of water, coal fines, silt, sand, and clay particles, which is most commonly disposed of in an impoundment.

COAL REFUSE IMPOUNDMENTS IN THE UNITED STATES

Coal refuse disposal impoundments are constructed for the permanent disposal of any coal, rock, and related material removed from a coal mine in the process of mining. Standard classification of coal slurry impoundments includes the following:

- *Active*—In operation and receiving slurry.
- *Inactive*—Not in operation or receiving slurry. Inactive impoundments may receive slurry in the future, becoming active again, and therefore have not been closed permanently.
- *Abandoned*—Not in operation and closed. These impoundments usually have been filled to capacity and have been closed and reclaimed.
- *Grandfathered or "Pre-law"*—Not in operation since promulgation of the 1977 Surface Mining Control and Reclamation Act (SMCRA) regulation. These impoundments are reclaimed under the Abandoned Mine Lands Program.

As of August 2001, MSHA oversees 713 active fresh-water and slurry impoundments in the United States (T.Bentley, Mine Safety and Health Administration, personal communication, 2001). Most coal waste impoundments in the United States are found in the East, predominantly in West Virginia, Pennsylvania, Kentucky, and Virginia. The thicker Western coal seams being mined now contain fewer in-seam partings and out-of-seam rock, and most of that coal is shipped raw without extensive processing. Consequently, coal waste impoundments are rarely used in the West. In the Eastern coalfields, however, the majority of the coal from underground mines is processed before sale. Of the 1.1 billion tons of coal mined in the United States per year, about 600–650 million tons are processed to varying degrees. Typically 350–400 million tons are handled in wet-processing systems that, on average, produce 70–90 million tons of fine refuse in a slurry. (C.Raleigh, CQ Inc., personal communication, 2001). Much of that material is stored in coal waste impoundments.

Disposal methods for coarse and fine coal refuse developed along separate lines. Even before the days of modern coal preparation plants, laborers picked rock from the run-of-mine coal and discarded it in refuse piles. Sometimes coarse refuse was returned to the mine, but more commonly it was deposited on the surface.

When fine coal cleaning came into widespread use, it became necessary to deal with the refuse. One way is to pump the slurry into an impoundment and allow the particles to settle. Another is to concentrate or dewater the slurry and/or to mix it with coarse refuse or other additives (e.g., lime, sodium silicate, elastomeric polymers, resinous adhesives) to provide stability (Osborne, 1988), and then dispose of it in a landfill or impoundment.

DISPOSAL OF FINE REFUSE IN IMPOUNDMENTS

To impound fine coal slurry, embankments are constructed with compacted coarse coal refuse material. Prior to an accident in 1972, the Buffalo Creek disaster (Sidebar 1.3), little consideration was given to control of water entering an impoundment from a preparation plant or as runoff from the watershed above. Indeed, the coarse coal refuse used for embankment construction provides a filter to limit impacts to the quality of the water entering nearby streams (D'Appolonia Consulting Engineers, 1975). In most impoundments, the embankment is constructed of coarse coal refuse, according to a design that is approved by regulatory authorities (see Chapter 2).

In the mountainous Appalachian region, the coarse refuse embankments are usually constructed across a valley, enclosing a basin that holds the coal refuse. In flatter Midwestern terrain, bermed and incised impoundments may be constructed. They typically have a larger surface area and are shallower. The slurry is pumped into the impoundment, where the fine particles in the slurry settle by gravity beneath a pool of clear water. In many impoundments, this clear water is pumped back to the coal processing plant and is reused.

SIDEBAR 1.3 FEBRUARY 26, 1972: BUFFALO MINING COMPANY, BUFFALO CREEK, WEST VIRGINIA

On February 26, 1972, the most destructive flood in West Virginia's history occurred when a coal waste impounding structure collapsed on the Buffalo Creek tributary of Middle Fork. Shortly before 8:00 a.m., the impounding structure collapsed, releasing approximately 132 million gallons of water. The water passed through two more piles of coal waste blocking the Middle Fork. At that time, there were no federal standards requiring either impoundments or hazardous refuse piles to be constructed and maintained in an approved manner.

Around 1957, as part of its surface mining operations, the Buffalo Mining Company (a subsidiary of the Pittston Coal Company) had begun depositing mine waste consisting of rock and coal in Middle Fork. Buffalo Mining constructed its first impounding structure, near the mouth of Middle Fork in 1960. Six years later, it added a second impounding structure, 600 feet upstream. By 1968, the company was depositing more waste another 600 feet upstream. By 1972, the height of this third impounding structure ranged from 45 to 60 feet.

Between February 24 and 26, 1972, the National Weather Service measured 3.7 inches of rain in the area of Logan County and Buffalo Creek. The impounding structure probably failed because foundation deficiencies led to sliding and slumping of the front face of the refuse bank. The waterlogged refuse bank accelerated the failure. The slumping lowered the top of the refuse bank and allowed the impounded water to breach and then rapidly erode the crest of the bank. Upon failure of the refuse bank, the floodwater moved into pockets of burning coal waste.

As result of the flood, 125 people were killed, 1,100 were injured, and more than 4,000 were left homeless. In addition, the flood completely demolished 1,000 cars and trucks, 502 houses, and 44 mobile homes, and damaged 943 houses and mobile homes to varying extents. Property damage was estimated at $50 million.

SOURCE: W.E.Davies et al., 1972

FINES DISPOSAL PROBLEMS IN OTHER MINING SECTORS

The problem of slurry disposal is not unique to the coal industry; it is a consideration for many base and precious metals industries, as well. For example, in the aluminum industry, disposal of massive quantities of bauxite tailings (called red mud) creates similar problems (Wagh and Desai, 1987). Disposing of the red mud in settling ponds in dilute, fine mud-sand slurries of about 20 percent solids (Downs and Stocks, 1977) brings with it a number of problems including very slow settling time and low bearing strength. The search for alternative disposal technologies (e.g., thickened [60 percent solids] high-density residue stacking), as well as for processing and utilization of the red mud, have been underway for some time (Wagh and Desai, 1987).

The disposal of the tremendous volume of waste sand and clay in the beneficiation process of phosphate mining is another example of these problems. The clay is pumped to a settling area in a dilute stream (3 percent solids) or sent to a thickener to increase the solids content (15 percent solids) (Garlanger and Fuleihan, 1983). Fine sediment disposal also is a problem in base metal mining and smelting operations.

COAL WASTE IMPOUNDMENT FAILURES

Coal waste facilities have been involved in several accidents or incidents since 1972. The incidents reviewed here demonstrate the range of the types of failures that can affect coal waste impoundments and of impacts of such failures. They are not a complete list of incidents. The first event was the Buffalo Creek accident (Sidebar 1.3), the most serious because it resulted in the loss of 125 lives and extensive damage to property downstream of the refuse piles and impoundments. Following that accident, regulations were promulgated to govern the design of the embankment structures used in future impoundments. Since then, no engineered embankments have failed, although other incidents and accidents have occurred. Sidebars 1.4 to 1.12 describe selected events.

SIDEBAR 1.4 AUGUST 14, 1977: ISLAND CREEK COAL COMPANY, BOONE COUNTY, WEST VIRGINIA

An embankment under construction failed at Island Creek Coal Company's impoundment in Boone County, West Virginia, on August 14, 1977. Heavy rainfall overflowed a temporary diversion ditch, causing the water level in the impoundment to rise. Because the embankment was still under construction, storage capacity had not yet reached the required minimum, and the sudden influx of additional water overtopped the embankment. Meanwhile, the water eroded the embankment, reducing its height 23 feet during a two-day period. During this time, 6.8 acre-feet of material was released, which clogged a drainage pipe downstream.

SOURCE: Owens, 1977.

SIDEBAR 1.5 DECEMBER 18, 1981: EASTOVER MINING COMPANY, HARLAN COUNTY, KENTUCKY

On December 18, 1981, Eastover Mining Company's Hollow No. 3 combined refuse disposal site failed, releasing about 25 million gallons of saturated coal refuse. The operation had been permitted to dispose of layers of coarse coal refuse and dewatered slurry "filter cake," which contained approximately 30 percent moisture, behind an embankment (see discussion of disposal techniques in Chapter 7) and, at a height of 192 feet, had reached 90 percent of its planned capacity. Several factors contributed to the increased pore water pressure in the dewatered fine refuse zone, including: (1) the filter cake layers had not been allowed sufficient time to dry before additional material was added; (2) layers of filter cake were not completely covered with coarse coal refuse; (3) a stream flowed into the impounded material, increasing saturation; and (4) material used in construction of the embankment did not allow water to seep out. The failure released a mudflow approximately 5 feet deep that traveled 4,400 feet downstream (500 feet in vertical distance) into the community of Ages, Kentucky. One resident was killed, three houses were destroyed, and 30 homes were damaged.

SOURCE: Cannon, 1981.

SIDEBAR 1.6 APRIL 8, 1987: PEABODY COAL COMPANY, RALEIGH COUNTY, WEST VIRGINIA

On April 8, 1987, a breach developed in the principal spillway pipe in the Lower Big Branch impoundment at Peabody's Montcoal No. 7 complex in Raleigh County, West Virginia. The 36-inch-diameter pipe ran through the impoundment and under part of the embankment at a depth of 55 feet. The rupture released nearly 23 million gallons of water, slurry, and fine coal refuse.

The exact cause of the accident was not identified but was probably a combination of factors: (1) Heavy snowfall (16 inches of snow with a rainfall equivalent of 1.9 inches), followed by rapid temperature increases and snowmelt, sent excessive amounts of water through the pipe. (2) Two landslides occurred in the slope above the rupture. Although the relative timing of the landslides and the breach is not known, the slides could have caused the pipe to collapse or separate. (3) Erosion of particles near the pipe connections could have reduced the bearing strength of the pipe. (4) The strength of an "elbow" in the piping may have been exceeded by massive and rapid fluid flow. In addition, a sinkhole that developed from the rupture threatened the stability of the embankment. The sinkhole came within 100 feet of several upstream-constructed additions to the cross-valley embankment before stability was maintained through mitigation of the breach.

The impoundment, upstream from several communities, was rated at the time as high hazard. A 50-mile stretch of Coal River from Montcoal to its mouth at St. Albans was visibly affected, and five water plants were shut down. Although 1,700 customers' water supply was disrupted in the Racine Public Service District, no human injuries or fatalities occurred as a result of this incident.

SOURCE: Owens, 1987.

SIDEBAR 1.7 JANUARY 28, 1994: CONSOLIDATION COAL COMPANY, MORGANTOWN, WEST VIRGINIA

On January 28, 1994, a 5-foot earthen berm failed at a slurry refuse impoundment at the Arkwright Mine in Granville, West Virginia. Heavy rain and melting snow resulted in 30 inches of water collecting behind the berm; it was determined that the 4-inch discharge pipe and rock underdrain at the site were insufficient to prevent water accumulation. The incident released 375,000 gallons of water into the town of Granville. Although no one was injured, three residences directly downstream were damaged.

SOURCE: Betoney, 1994.

SIDEBAR 1.8 MAY 22, 1994: MARTIN COUNTY COAL CORPORATION, DAVELLA, KENTUCKY

On May 22, 1994, a breakthrough occurred at Martin County Coal Corporation's Big Hollow slurry impoundment in Davella, Kentucky. Nearly 32 million gallons of black water inundated an abandoned and sealed-off portion of the mine. The breakthrough resulted either from collapse or water penetration of the Coalburg coal seam bordering the impoundment. Slurry had been impounded 32 feet higher than the coal seam's elevation. The mine's 16-inch concrete-block seals held the black water inundating the mine, but water broke through portal seals and a coal seam outcrop barrier. Although the slurry level dropped by 6 feet, the embankment structure was not damaged, and no injuries or fatalities occurred.

SOURCE: Stewart and Robinson, 1994.

SIDEBAR 1.9 AUGUST 9, 1996: LONE MOUNTAIN PROCESSING INCORPORATED, ST. CHARLES, VIRGINIA

On August 9, 1996, there was a breakthrough at Lone Mountain Processing's Miller Cove slurry impoundment. The evening before the failure, approximately 2.75 inches of rain had fallen, and most of it within an hour and a half. Approximately 1 million gallons of black water were released into Gin Creek through an abandoned mine. (Underground mines had operated in areas adjacent to the impoundment from the 1920s to the 1980s.)

Excavation of the breach showed that the leak occurred in an area where available mine maps indicated a barrier of at least 25 feet of solid coal between the outcrop and the underground mine workings. Further exploration revealed that the barrier was in fact less than 2 feet thick. It is believed that hydrostatic pressure from the slurry opened cracks in the coal seam and began a piping-type failure. The thin coal barrier was progressively eroded, allowing slurry to flow uncontrolled into the abandoned mine.

SOURCE: Michalek et al., 1996.

SIDEBAR 1.10 OCTOBER 24, 1996: LONE MOUNTAIN PROCESSING INCORPORATED, ST. CHARLES, VIRGINIA

On October 24, 1996, a second breakthrough occurred at Lone Mountain Processing's Miller Cove impoundment, but in another area of the abandoned mine. This release was more serious than the event in August (Sidebar 1.9) because the water contained more solids. Approximately 6 million gallons of water and slurry exited the abandoned mine into Gin Creek and flowed 11 miles, where it entered the Powell River's North Fork. Reportedly, the river was discolored for more than 40 miles.

The failure resulted from two large sinkholes that had developed on the northwestern end of the impoundment. When the site was excavated to locate the breach, it was determined that the slurry had entered through a fracture in the mine roof that coincided with these sinkholes.

SOURCE: Michalek et al., 1996.

SIDEBAR 1.11 NOVEMBER 26, 1996: CONSOLIDATION COAL COMPANY, OAKWOOD, VIRGINIA

On November 26, 1996, the Buchanan No.1 impoundment in Buchanan County, Virginia, failed. In the 1960s, the Kennedy coal seam at the site had been excavated by both surface area mining and underground auger mining. After the impoundment was constructed (1984), another company mining underground in the adjacent drainage area apparently intersected the historic auger mine workings, providing a conduit for the slurry.

Coal refuse and slurry from the impoundment broke into an abandoned underground mine and discharged about 1,000 gallons per minute at its peak through two mine portals into the adjacent North Branch Hollow of the Levisa Fork of the Big Sandy River. There was no detrimental impact on the embankment, and no one was killed or injured.

SOURCE: Michalek et al., 1996.

SIDEBAR 1.12 OCTOBER 11, 2000: MARTIN COUNTY COAL CORPORATION, INEZ, KENTUCKY

On October 11, 2000, a coal waste impoundment of the Martin County Coal's preparation plant near Inez, Kentucky, released slurry containing an estimated 250 million gallons of water and 31 million gallons of coal waste into local streams. Reportedly, the failure was caused by the collapse of the slurry pond into underground coal mine workings next to the impoundment. The slurry broke through an underground mine seal and discharged from mine entrances 2 miles apart into two different watersheds (Wolf Creek and Coldwater Fork).

Although no human life was lost, the release killed aquatic life along the Tug Fork of the Big Sandy River and its tributaries. Public water supplies were disrupted when communities along the rivers in both Kentucky and West Virginia shut down water plants to prevent contamination with black water. American Electric Power had to close its massive generating plant, and numerous properties and residences were damaged.

SOURCE: Various issues of the *Herald Leader,* the *Courier-Journal,* and the *Charleston Gazette* (2000, 2001); Dennis Hatfield, Martin County Coal Corporation, personal communication, 2001.

Two of the events resulted from leaks or failures of drainage pipes. However, the majority of the incidents involved failures in the basin area. Inaccurate mine maps and inadequate characterization of the basin area most likely contributed to at least some of these incidents. This report deals with both of these issues in detail (Chapters 4 and 5).

IMPOUNDMENT HAZARD RANKING SYSTEMS

MSHA bases its hazard potential rating system on the height of the embankment, the volume of material impounded, and the downstream effects of an impoundment failure (MSHA, 1974, 1983). The resulting three classifications are:

- *Low Hazard Potential*—Facilities in rural areas where failure would cause only slight damage, such as to farm buildings, forest, agricultural land, or minor roads.
- *Moderate Hazard Potential*—Facilities in predominately rural areas where failure may damage isolated homes or minor railroads, disrupting services or important facilities.
- *High Hazard Potential*—Facilities whose failure could reasonably be expected to cause loss of human life, serious damage to houses, industrial and commercial buildings, important utilities, highways, and railroads.

The MSHA guidelines indicate that design criteria become more conservative as the hazard potential increases. For example, design criteria for the maximum precipitation (flood) event increase as the hazard classification moves from low to high. Thus, storm design criteria for a long-term high-hazard-potential impoundment require that the impoundment be designed to contain the probable maximum precipitation that is reasonable for the region (MSHA 1974, 1983). In addition, piezometers are generally required to monitor and verify the water saturation conditions within the embankment for moderate and high hazard sites.

MSHA guidelines further state that the stability of an embankment should normally have minimum static and seismic factors of safety of at least 1.5 and 1.2, respectively, under maximum anticipated phreatic conditions. The guidelines require extra attention to seismic events for high hazard impoundments in certain regions (MSHA, 1974, 1983).

On December 1, 1997, after two unintentional releases of slurry in Virginia in a two-month period, MSHA introduced a second classification system that addresses the potential for the unintentional release of water or slurry from impoundments into active or abandoned mines (Sidebars 1.9 and 1.10). This classification system allows the coal mine operator to evaluate the proximity of underground workings; to determine whether the recommendations of Bureau of Mines Information Circular 8741 (Babcock and Hooker, 1977), which provides guidelines for mining under bodies of water, are met; and to assess the potential impact if a breakthrough were to occur. For example, the impact may threaten the safety of miners or the safety of the general public (MSHA, 1997). A priority rating is assigned to each impoundment based on its breakthrough potential—whether low, medium, or high—and a potential impact of a breakthrough. The purpose of this classification system is to evaluate whether the impoundment plan adequately addresses the breakthrough potential (MSHA, 1997). These ranking systems are based on the proximity of the basin to underground workings as well as the potential downstream impacts were a basin to fail. However, they do not assess the probability of failure. It is a completely separate ranking than that which is done for the embankment structure.

In addition, the Federal Emergency Management Agency ranks embankments based on potential impacts should a failure occur (U.S. Bureau of Reclamation, 1988). This inventory lists more than 76,000 dams. The Office of Surface Mines (OSM) and the state delegate programs use a similar system to rank earth dams and reservoirs by whether they are located in rural areas and the amount of damage failure could cause (U.S. Department of Agriculture, 1976). Neither of these organizations has a ranking system for breakthrough potential.

STUDY AND REPORT

Concern about the potential for accidents like the one at Inez, Kentucky (October 2000), motivated Congress to direct MSHA to commission an independent study of current coal waste disposal methods and an exploration of alternatives for future disposal of coal waste. In addition, Congress directed that the study examine engineering standards for coal waste impoundments, and recommend ways to improve the stabilization of impoundment structures.

The National Research Council (NRC) established the Committee on Coal Waste Impoundments to undertake this study. The committee consists of 14 experts from academia, industry, and state government with expertise in coal mining, geology, geophysics, geochemistry, hydrology, mining regulations, environmental law, mining health and safety, land-use planning, and geotechnical and geological engineering. Brief biographies of the committee members appear in Appendix A.

The overall objectives of this study are:

- to examine engineering standards for coal waste impoundments;
- to provide recommendations for improving impoundment structure stabilization;
- to determine the adequacy of mine maps; and
- to evaluate potential alternatives for future coal waste disposal, including the benefits of each alternative.

The Statement of Task lists the following specific tasks:
Engineering Standards/Barrier Stability/Monitoring

- Examine current engineering practices for coal waste impoundments and provide recommendations for improving engineering practices, including impoundment structure stabilization.
- What alternative means are available to evaluate or confirm the safety of the designed barrier(s) protecting slurry impoundments?
- What options can be developed to effectively monitor the status of coal barriers left to protect slurry impoundments?

Site Characterization

- Evaluate the adequacy of mine maps and explore ways to improve mapping and surveying practices in general for the mining industry.
- What is the best way to three-dimensionally conceptualize and delineate the impoundment area, including the extent of underground mine works beneath or adjacent to the slurry disposal area?

Alternative Technologies

- Evaluate potential alternatives for future coal waste disposal, including the benefits of each alternative.
- Are there other methods to wash and process coal that would reduce the amount of slurry disposal needed?
- What are the options for the coal waste product to be refined further in order to produce a marketable product?

To address the charge, the committee gathered, synthesized, and analyzed information by working in subgroups based on the three main topics in the Statement of Task (geotechnical aspects, site characterization, and alternative technologies). The committee held eight information-gathering meetings, including six subgroup meetings, between March and June 2001. The meetings included presentations by, and discussions with the sponsor, personnel from other government programs, and representatives of industry, academia, environmental organizations, labor organizations, and citizens' groups (Appendix B). To obtain input from the public, the committee held eight town meetings throughout the Eastern coal-mining region. Finally, the full committee met twice in closed session for discussion and writing. As background material, the committee reviewed relevant government documents and materials, pertinent NRC reports, and other technical reports and literature published through July 2001.

This report is intended for multiple audiences. It contains advice for MSHA, OSM, other federal agencies, and state regulatory agencies, as well as policy makers, the coal industry and its consultants, scientists, and engineers. Chapter 2 gives an overview of the regulatory framework for coal waste impoundments. Chapter 3 examines issues related to the engineering design for coal refuse facilities. Chapter 4 discusses site characterization, including mine mapping, map storage and preservation, and surveying. Chapter 5 addresses techniques, such as geophysical methods, to locate abandoned mines and other voids, and hydraulic testing to establish the thickness of barrier pillars. Chapter 6 discusses ways to limit potential failure modes for the embankment and basin area. Chapter 7 addresses alternatives to slurry impoundments, including alternative mining and coal preparation methods, direct utilization of slurry, and alternative disposal techniques. Chapter 8 summarizes the committee's conclusions and recommendations. Technical terms are defined in the glossary (Appendix C).

It is important to recognize that this charge specifically directs the committee to focus its analysis on the engineering and characterization of coal waste impoundments. The committee was not asked to consider other factors related to potential impacts of disposing of coal wastes in an impoundment, or any other disposal option. For example, these factors might include potential long-term effects on water quality; land use issues, including long-term stewardship of closed impoundments; and economic and cost-benefit analyses of alternatives. The committee also was not asked to evaluate the risks of individual impoundments, examine the qualifications and training of inspectors, or comment on coal mining policy issues not directly related to impoundments. Although important, such issues are well beyond the charge to this committee. Furthermore, a comprehensive analysis of these issues would require considerably more time than was available for the present study.

2

Current Regulatory Framework

Numerous federal and state statutes and regulations apply to coal mining activities, including disposal of coal waste in impoundments. Some of these statutes and regulations are specific to coal mining, such as the Federal Mine Safety and Health Act (1977) and the Surface Mining and Reclamation Act of 1977 (SMCRA), and regulations adopted thereunder. Coal mining activities and facilities are subject to more widely applicable statutes and regulations, such as the Clean Air Act and the Clean Water Act and regulations adopted thereunder.

Nearly every coal waste facility is subject to regulatory requirements imposed by MSHA and either OSM or a state with a regulatory program approved under SMCRA. Because state program organization varies from state to state, and because some states opt not to obtain primacy, this chapter focuses on federal statutory provisions and MSHA and OSM regulations that directly relate to the design, construction, operation, and closure of refuse impoundments, as well as alternative refuse disposal techniques. Other federal statutes that may relate to refuse impoundments and refuse disposal, including the Clean Air (1970), the Clean Water Act (1977), the Safe Drinking Water Act (1974), the Resource Conservation and Recovery Act (1976), the Migratory Bird Treaty Act (1918), and the Endangered Species Act (1973), are discussed in less detail. These statutes involve federal agencies, like the U.S. Environmental Protection Agency (EPA), the U.S. Army Corps of Engineers, the U.S. Fish and Wildlife Service, as well as state agencies with an implementation role under those statutes.

FEDERAL MINE SAFETY AND HEALTH ACT

Congress specifically stated that the purpose of the Federal Mine Safety and Health Act of 1977 was to establish mandatory standards to protect the health and safety of miners (30 U.S.C. [United States Code] § 801 (a) and (g)). This limitation is important in assessing MSHA's regulatory responsibility and authority associated with potential impoundment breakthroughs and failures that do not affect the health and safety of miners but may affect persons, communities, or ecosystems downstream.

MSHA developed standards for impoundments and refuse piles after the Buffalo Creek disaster in February 1972 (Sidebar 1.3) and promulgated regulations to address the design, review, and monitoring of these structures (30 C.F.R. [Code of Federal Regulations] §§ 77.214, 77.215, and 77.216). While these regulations address a variety of health and safety issues pertaining to all aspects of coal mining, Section 77.216 is specifically directed at water, sediment, and slurry impoundments and impounding structures. It defines the identification, engineering plan, inspection, hazard abatement, reporting, certification, and abandonment requirements for impoundment systems that fall under MSHA's jurisdiction. Impoundments that are large enough to be regulated by MSHA are commonly referred to as MSHA-class impoundments (30 C.F.R. § 77.216). These structures impound water, sediment, or slurry to an elevation of 5 feet or more and a storage volume of 20 acre-feet or more, or impound to an elevation of 20 feet or more, or present a hazard to coal miners, as determined by the local district manager. The following discussion, while not a comprehensive overview of the regulatory requirements of 30 C.F.R. § 77.216, presents the salient points of the MSHA regulations that are relevant to limiting the potential for refuse impoundment breakthroughs and failures.

Requirements for Engineering Plans

Regulations contained in 30 C.F.R. § 77.216 require that plans for the design, construction, and maintenance of MSHA-class impoundments be prepared and submitted to MSHA for approval before construction begins. Such plans must include information on the physical and engineering properties of foundation materials and of embankment construction materials. They must include a stability analysis of the impounding structure and identify the location of surface and underground mine workings near the facility. Specific to underground workings, 30 C.F.R. § 77.216–2 (a)(14) requires that:

> The locations of surface and underground coal mine workings including the depth and extent of such workings within the area 500 feet around the perimeter, [should be] shown at a scale not to exceed one inch=500 feet.

A registered professional engineer must certify that the impounding structure has been designed in accordance with prudent engineering practice.

Review and Approval Process

MSHA is subdivided into 11 districts, each under the direction of a district manager who is responsible for a given geographic region of the United States (Figure 2.1). The districts have multiple offices to accommodate the day-to-day process of inspecting mines. Approval of engineering plans and inspections is carried out by the district office staff. Most of the professional employees in these offices are inspectors who report to the district manager through an intermediate supervisor and who regularly visit the mines in the area.

The MSHA Mine Waste and Geotechnical Engineering Division in Pittsburgh, Pennsylvania has also been established within the agency structure to provide districts with technical expertise in their review of the engineering plans. This technical team acts in a consulting capacity to the district offices. When a design for a new coal refuse impoundment, or a modification to an existing impoundment, is submitted to the district office, the district manager determines whether to approve the submission, ask for more information, or ask for assistance from the technical team. In most instances, the district manager asks the technical team for assistance. The technical team has no approval authority and acts only in the capacity of a technical consultant to the district manager. Facility construction or modification can only commence after the district manager has approved the plan.

The technical team reviews the engineering plans, designs, and data submitted by the professional engineer. The primary focus of the review is the stability of the impounding structure. If the technical team cannot reach a decision based on the information submitted, it asks for additional information. Typically, several exchanges occur between the technical team and the design engineer to ensure that the engineering plans are both complete and technically sound. When the technical team is satisfied with the engineering design of an impoundment, it recommends that the district manager approve the plan.

Because MSHA's primary concern is the welfare of miners, MSHA's focus for underground mining with regard to impoundments is the effects of mining on embankment stability, and the possibility of water inrush into an underground mine or onto the surface area of mines. To deal with these concerns, MSHA relies on the operator's determination of the presence of

FIGURE 2.1 Coal Mine Safety and Health Administration district offices. Modified from data provided by MSHA.

underground workings. The review procedures provided to this committee did not indicate any independent determination of the depth and extent of underground workings.

Inspection Requirements

To be approved, a construction plan must contain a plan for construction surveillance, maintenance, and repair of the impounding structure; monitoring of the disposal process is mandatory (30 C.F.R. § 77.216–2 (a)(15)). Compaction of the material is the primary concern, including the method of emplacing the material, its size and consistency, and thickness of the layers. This specific monitoring of construction is unique to each impoundment and detailed in the approved plan. All MSHA-class water, sediment, and slurry impoundments must be inspected at least once every seven days by an MSHA-certified, "qualified person." The MSHA policy manual states that this person "must be…trained according to an approved 30 C.F.R. § 77.216 training plan that has a program of instruction on impoundment inspection, …pass an evaluation, [and may be required to demonstrate successful training by] oral, written, or practical demonstration."

The MSHA regulations for monitoring the stability of an embankment are comprehensive. If the MSHA regulations for monitoring are met, then regulations administered by the state are usually met as well, except that MSHA does not require monitoring of discharge points for water quality. The MSHA requirements for monitoring are specific, namely:

- The impoundment must be larger than a minimum size.
- A qualified person must make the examination.
- The examination must be made at intervals not to exceed seven days.
- Observations are required for the appearance of structural weakness and other hazardous conditions.
- All instruments shall be monitored during this examination.
- The results of the examination are recorded and reviewed by one of the officials in charge of the mine.

The MSHA regulations require impoundment operators to develop specific plans and procedures for dealing with hazardous or emergency situations, which includes notifying the district manager. If a hazardous condition is detected, the monitoring interval is reduced to at least once each eight hours until the hazard abates.

Monitoring requirements are designed to ascertain the stability of the embankment. Examiners look for cracks, seeps, or slumping material on the slopes. The instrumentation usually measures the extent of water seepage within the embankment (higher levels can be destabilizing), and sometimes instruments are installed to detect movement within the body of the embankment. Another requirement during the examination is that the maintenance and conditions of the principal and emergency spillways be checked. In most instances, the required examination can be completed at or near the embankment.

The regulations do not cover several aspects of monitoring impoundment construction and operation. Even though the depth, thickness, and structure of the layers of rock and soil between the impoundment basin and a mined coal seam are addressed in the plan approval process, no specific monitoring requirements focus on this subject. In addition, the monitoring regulations do not require the inspector to address mine openings to the surface from underground workings under the impoundment, nor bulkheads that might be installed within those openings. MSHA may require such monitoring in individual cases, but there is no regulatory requirement that it do so.

In addition to the monitoring required of the impoundment operator, inspectors from MSHA and state agencies regularly inspect the impoundment. The agency inspections satisfy regulatory or policy-required inspection schedules and have no effect on monitoring requirements the regulations impose on the operator. The Federal Mine Safety and Health Act requires MSHA to inspect each impoundment at least twice each year. (If the impoundment is associated with an underground mine it must be inspected four times a year.) In addition, qualified persons may be assigned to inspect the impoundment more often if MSHA deems it necessary.

Reporting Obligations

A report of operations documenting the overall condition and performance of the refuse impoundment must be submitted to MSHA after construction begins. The district manager, in the exercise of his discretion, determines the appropriate frequency case by case. A registered professional engineer must certify these annual reports, attesting that work accomplished at the facility during the past 12 months has been done according to the approved plan (30 C.F.R. § 77.216–4).

Observations

The regulatory language in 30 C.F.R. § 77.216 speaks almost exclusively to the engineering aspects and stability of the embankment. There is little reference to the engineering design of the basin area or to areas other than the embankment of the impounding structure.

The impoundment operator and professional engineer are responsible for providing information about underground mine workings, including the depth and extent of the workings. However, no regulation or standard industry practice instructs or guides the engineer in this task, nor is there a procedure for an independent verification of the information submitted.

In addition, no specific regulation requires an evaluation of the breakthrough potential of impoundments. However, in practice, MSHA appears to have the jurisdiction to require an evaluation of breakthrough potential through indirect regulatory language. The regulations require that the engineering plans contain "other information pertinent to the impoundment itself, including any identifiable natural or manmade features which could affect operation of the impoundment" (30 C.F.R. § 77.216–2(7)), and that MSHA be presented with "other such information pertaining to the stability of the impoundment and impounding structure which may be required" (30 C.F.R. § 77.216–2(18)). While neither of these requirements specifically obliges consideration of the potential interactions between the surface impoundment and underground mine workings, the language does allow MSHA the ability to require such evaluations.

SURFACE MINING CONTROL AND RECLAMATION ACT

SMCRA (30 U.S.C. §§ 1201 to 1328) established OSM within the Department of the Interior and assigned the agency authority to implement a national program to regulate the surface effects of coal mining activities. Congressional intent in adopting SMCRA was to protect "society and the environment from the adverse effects of surface coal mining operations" (30 U.S.C. § 1202 (a)). This legislative and regulatory direction complements the intent of the Federal Mine Safety and Health Act, which provides for the health and safety of miners, by adding the SMCRA directive to protect the public and to limit environmental effects. The program authority has been delegated to the states to perform the administrative and enforcement functions, but OSM retains oversight of program adequacy and administers the program in states without primacy.

SMCRA specifies that, because of the diversity in terrain, climate, biology, geochemistry, and other physical conditions under which mining operations occur, the primary governmental responsibility for regulating surface coal mining and reclamation operations should rest with the states (30 U.S.C. § 1201 (f)). To achieve primary regulatory responsibility, often referred to as primacy, a state must develop a program that demonstrates the state's capability to carry out the relevant provisions of SMCRA (30 U.S.C. § 1253). Once the Secretary of the Interior approves a state regulatory program, OSM assumes an oversight role, making inspections as necessary to evaluate the state's administration of their program and taking corrective action when necessary.

Currently, 24 states have primacy: Alabama, Alaska, Arkansas, Colorado, Illinois, Indiana, Iowa, Kansas, Kentucky, Louisiana, Maryland, Mississippi, Missouri, Montana, New Mexico, North Dakota, Ohio, Oklahoma, Pennsylvania, Texas, Utah, Virginia, West Virginia, and Wyoming. Most of the regulatory requirements of these state programs are similar to the OSM regulations at 30 C.F.R. Chapter VII, but, in response to local concerns and conditions, some programs contain requirements that are more extensive than the federal rules.

OSM directly regulates surface coal mining and reclamation operations on all Indian lands and in nonprimacy states. Currently 12 states do not have a primacy program (Arizona, California, Georgia, Idaho, Maine, Michigan, North Carolina, Oregon, Rhode Island, South Dakota, Tennessee, and Washington). Of those, only Tennessee, Washington, and Arizona, have significant active surface coal mining operations.

Because the OSM regulations in 30 C.F.R. Chapter VII form the template for state programs, this chapter focuses on the federal regulations rather than the requirements of individual state regulatory programs. Several sections of the regulations are specific to the design, construction, and maintenance of coal refuse impoundments. Other sections apply in a more general fashion. The following summary highlights requirements relevant to limiting the potential for impoundment breakthroughs and failures.

Buffer Zone Rule

Many coal waste impoundments in the Appalachian coalfields are placed in valleys and can affect the preexisting drainage. SMCRA, as well as parallel state laws, governs the placement of mining materials in or near a stream. SMCRA at 30 U.S.C. § 1265(b) contains general environmental performance standards for protecting streams and fish and wildlife. To carry out these and other requirements of the statutory program, the OSM promulgated a federal buffer zone rule in 1981, which provides (30 C.F.R. § 816.57):

> (a) No land within 100 feet of a perennial stream or an intermittent stream shall be disturbed by surface mining activities, unless the regulatory authority specifically authorizes surface mining activities closer to, or through, such a stream. The regulatory authority may authorize such activities only upon finding that: (1) Surface mining activities will not cause or contribute to the violation of applicable State or Federal water quality standards, and will not adversely affect the water quantity and quality or other environmental resources of the stream.

Federal SMCRA regulations at 30 C.F.R. § 701.5 include the following stream-related definitions:

> *Intermittent stream means* (a) a stream or reach of a stream that drains a watershed of at least one square mile, or (b) a stream or reach of a stream that is

below the local water table for at least some part of the year, and obtains its flow from both surface runoff and groundwater discharge.

Perennial stream means a stream or part of a stream that flows continuously during all of the calendar year as a result of groundwater discharge or surface runoff.

Operators permitting a coal waste impoundment that lies within the buffer zone obtain waivers of this rule by obtaining Section 404 permits from the U.S. Army Corps of Engineers (see below).

Requirements for Engineering Plans

Impoundments that are constructed of, or that impound, coal waste materials must comply with the requirements of SMCRA at 30 U.S.C. §§ 1265 (b)(13), 1265(b)(f), and 1266(b)(5), and OSM regulations at 30 C.F.R. §§ 780.25, 816.49, 816.81 and 816.84. The rules require the submission of engineering plans for these impoundments. They also cross-reference the MSHA requirements for engineering plan content. They specify storm and freeboard design requirements, as well as minimum safety factors under reasonably foreseeable site conditions. Requirements for the embankment, foundation, and abutment conditions are addressed, as are seepage and allowance for settlement.

OSM regulations (30 C.F.R. § 816.81(d)) require that all applications for coal mine waste disposal facilities, including all coal waste impoundments, have a foundation investigation to determine design requirements for foundation stability. This rule specifically requires that the design include an analysis of foundation conditions that takes into consideration the effect of underground mine workings on the stability of the disposal facility. The regulations at 30 C.F.R. § 780.25 (e) also require that each design plan submitted for this type of facility include the results of a geotechnical investigation of the embankment foundation area. Finally, under 30 C.F.R. §780.25 (a)(1)(iv), each impoundment plan must include a survey describing the potential effect on the structure from subsidence of the subsurface strata resulting from past underground mining operations if underground mining has occurred.

Review and Approval Process

The content of the permit applications for impoundments is specified in 30 C.F.R. § 780.25. When a design for a new coal refuse impoundment, or a modification to an existing impoundment, is submitted to the SMCRA regulatory authority, the plans are reviewed by a geologist or engineer. The reviewers determine whether the application is complete in responding to all applicable regulatory requirements, whether the information is technically sound, and whether the proposed design and supporting documentation is adequate to support the findings and conclusions that the regulatory authority must make before approving a permit application. Mine operators must obtain all necessary permits, approvals, and authorizations from MSHA and the SMCRA regulatory authority before beginning to construct the facility. Changes in construction and operation after permit issuance, such as expansion, require permit revisions under 30 C.F.R. §774.15.

Inspection and Reporting Requirements

The OSM regulations (30 C.F.R. § 816.49(11)) require that a qualified, registered professional engineer, experienced in the construction of impoundments, inspect each coal waste impoundment regularly during construction, upon completion of construction, and at least annually thereafter until the impoundment is removed or until the site receives final bond release. After each inspection, the engineer must provide the regulatory authority with a certified report that the impoundment has been constructed and maintained as designed and in accordance with the approved plan and all applicable regulatory requirements. The report must include a discussion of any appearance of instability, structural weakness, or other hazardous condition, depth and elevation of any impounded waters, existing storage capacity, any existing or required monitoring procedures and instrumentation, and any other aspects of the structure affecting stability.

Under 30 C.F.R. § 816.49(a)(12), all impoundments that meet the size criteria of 30 C.F.R. § 77.216(a) also must be routinely inspected in accordance with the MSHA regulations in 30 C.F.R. § 77.216–3. Those impoundments that do not meet the criteria of 30 C.F.R. § 77.216(a) must be examined at least quarterly for the appearance of structural weakness and other hazardous conditions.

The states also employ qualified persons to inspect impoundments on varying schedules. Table 2.1 lists the monitoring requirements and inspection intervals for various states. The local OSM office, or state delegate

TABLE 2.1 Impoundment Monitoring Requirements by Selected States

State	Water Monitoring	Inspection Frequency	Professional Engineer Certification	Plan for Instrumentation	Emergency Warning Plans
Alabama	Yes	Two per month	Each construction phase	Yes	No[a]
Illinois	Yes	Quarterly for non-MSHA structures	At completion or quarterly	Yes	Yes
Indiana	Yes	Quarterly	Within 30 days of completion	Yes	Yes
Kentucky	Yes	As per MSHA	Quarterly, annually, at critical phase	Yes	No[a]
Maryland	Yes	Quarterly, annually, and per MSHA	Annually and at critical phase	Yes	No
Ohio	Yes	Monthly	Quarterly	Yes	Yes
Pennsylvania	Yes	Weekly	Critical phases and completion	Yes	Yes
Tennessee[b]	Yes	Monthly or quarterly	Quarterly and annually	Yes	No[a]
Virginina	Yes	Quarterly, critical phase and annually	Quarterly, critical phase, and annually	No	No[a]
West Virginia	Yes	Daily during construction	Quarterly	Yes	Yes

[a] While these states have no specific requirements, there is an MSHA requirement to create a plan.

[b] Tennessee does not have primacy. OSM oversees monitoring in this state.

SOURCE: Charles R.Gillian and Michael R.Wooldridge, Alliance Consulting, personal communication, 2001; Albert G.Morris, Indiana Department of Natural Resources, personal communication, 2001.

delegate program having primacy for mining, reclamation, and enforcement activities under OSM purview is responsible for regularly inspecting mines and responding to inquiries.

Observations

Like the MSHA regulations, the OSM regulations speak almost exclusively to the engineering aspects and stability of the embankment. The OSM regulations (30 C.F.R. §§ 816.49 and 816.84) mention no engineering, inspection, or reporting requirements for the impoundment basin area. These sections expressly pertain to structures that either are constructed of, or are built to impound, coal waste. While authority to require investigations of the impoundment basin area may be implied in general OSM regulations, since the basin is part of the impoundment, that authority is not explicit.

OTHER FEDERAL LAWS AND STATE DELEGATE PROGRAMS RELEVANT TO REFUSE DISPOSAL PRACTICES

Many federal and state laws may be applicable to coal mining activities and the operation of coal refuse disposal facilities. Significant federal laws relating to coal mining include the Clean Air Act (42 U.S.C. §§ 7401, *et seq.*), the Clean Water Act (33 U.S.C. §§ 1251, *et seq.*), the Safe Drinking Water Act (42 U.S.C. §§ 300f, *et seq.*), and the Resource Conservation and Recovery Act (42 U.S.C. §§ 6901, *et seq.*), where initial administrative and enforcement authority was assigned to the U.S. EPA. EPA may assign primary authority for its programs to the states provided that the states have established legal authority, a responsible agency, and an equivalent or more stringent program than that mandated by federal law and rulemaking. Because state program organization varies from state to state, and because some states opt not to obtain primacy, the following discussion focuses only on the federal requirements that must be carried through in a primary state program. These statutes are also implemented through detailed regulations found primarily in Titles 33 and 40 of the Code of Federal Regulations.

Clean Air Act

The Clean Air Act may require a coal operator to obtain a permit for particulate emissions released during coal preparation (42 U.S.C. § 7661a;

40 C.F.R. Part 60, Subpart Y). This requirement does not apply to refuse impoundments directly, but some states regulate fugitive dust emissions from these impoundments (e.g., Indiana, New Mexico). Certain alternative technologies to conventional refuse impoundment storage systems may require additional air permitting if the technology requires thermal processes such as synfuels production, fine coal drying, or mine-mouth fluidized-bed combustion.

Clean Water Act

Section 402 of the Clean Water Act (33 U.S.C. § 1342) establishes the National Pollutant Discharge Elimination System requirements for point-source discharges of pollutants to surface water. All releases of water through discrete conveyances from a coal mining operation must be permitted, monitored, and reported under the National Pollutant Discharge Elimination System program. Under this program, specific effluent limits define the maximum level of contaminants allowed in the discharge. EPA is currently in the process of applying watershed effluent limits based on total maximum daily load for certain contaminants to new and existing National Pollutant Discharge Elimination System permits (33 U.S.C. § 1313 (d)). Regulations for point-source discharges from coal mining are found at 40 C.F.R. Part 434.

The Clean Water Act, § 402, also obliges the EPA and its state delegate programs to protect the quality of surface water from sediment and other contaminants from nonpoint sources associated with surface runoff (40 C.F.R. § 122.26). Sedimentation ponds are constructed downgradient from the coal refuse impoundment to collect surface runoff from the embankment downslope and reduce sediment load before release. Releases from sedimentation ponds are subject to the National Pollutant Discharge Elimination System effluent discharge requirements, which contain limits for suspended solids.

Section 404 of the Clean Water Act (33 U.S.C. § 1344) authorizes the U.S. Army Corps of Engineers to permit the discharge of dredged or fill material into waters of the United States, including wetlands. Regulations promulgated by the U.S. Army Corps of Engineers include a definition of "fill" as follows (33 C.F.R. Part 323.2(e)):

> any material used for the primary purpose of replacing an aquatic area with dry land or of changing the bottom elevation of an [sic] waterbody. The term does not include any pollutant discharged into the water primarily to dispose of waste, as that activity is regulated under Section 402 of the Clean Water Act.

The U.S. Army Corps of Engineers has relied upon permit procedures in 33 C.F.R. Parts 320 and 323 and EPA guidelines at 40 C.F.R. Part 230 in their review of individual permits for valley fill, including coal waste impoundments, resulting from surface mining. The general standard is that:

> fill material should not be discharged into the aquatic ecosystem, unless it can be demonstrated that such a discharge will not have an unacceptable adverse impact either individually or in combination with known and/or probable impacts of other activities affecting the ecosystem of concern.

Section 404 permits are usually considered after the design reviews by MSHA and OSM (and their state counterparts) and require public review and comment opportunities. The consideration of a 404 permit by the U.S. Army Corps of Engineers may also require preparation of an Environmental Impact Statement under the National Environmental Policy Act. Because refuse impoundments in the Appalachian coalfields are often constructed in the upper reaches of a stream valley, a Section 404 permit is required for their construction, although whether waste materials can be permitted in a stream, under either a Section 404 or Section 402 permit, remains controversial (as discussed above).

The U.S. Army Corps of Engineers does not require an individual permit in all cases. They have adopted a Nationwide Permit Program, which authorizes a general permit designed to regulate with little, if any, delay or paperwork, certain activities having “minimal impacts,” including some coal waste impoundments in headwaters (stream tributaries where average annual flow is less the 5 cubic feet per second). If an impoundment qualifies under the Nationwide Permit, the applicants must notify the U.S. Army Corps of Engineers that they qualify and must show authorization by the OSM (or the state delegate program) and submit an approved mitigation plan. The Nationwide Permit contains a general provision for protecting the remaining segments of the stream.

Safe Drinking Water Act

The Safe Drinking Water Act establishes the Underground Injection Control program to protect underground sources of drinking water from effects associated with the injection of waste into the subsurface (42 U.S.C. §§ 300f, *et seq.*; 40 C.F.R. Part 144). The Underground Injection Control program is applicable to coal refuse disposal methods that inject refuse slurry back into underground mine workings. Although no regulations at the federal level require permits, Underground Injection Control programs may require permits at the state level.

Resource Conservation and Recovery Act

The Resource Conservation and Recovery Act regulates hazardous and nonhazardous solid wastes (42 U.S.C. §§ 6901, *et seq.*). Refuse generated by mining and beneficiation processes associated with mineral extraction, including coal, are categorically excluded from the definition of hazardous waste (42 U.S.C. § 6921 (b)(3)(A)(ii); 40 C.F.R. § 261.4(b)(7)). Other solid wastes, such as waste oils and solvents, fall within the definition.

Wildlife Protection Acts

The U.S. Fish and Wildlife Service is responsible for enforcing the Migratory Bird Treaty Act (16 U.S.C. §§ 703, *et seq.*), the Endangered Species Act (16 U.S.C. §§ 1531, *et seq.*), and other wildlife protection legislation. For coal refuse impoundments, the U.S. Fish and Wildlife Service is particularly interested in the influence of refuse disposal on riparian habitats, wetlands, and downstream aquatic communities. SMCRA and other statutes mandate coordination with the U.S. Fish and Wildlife Service in permitting activities that might affect wildlife. The U.S. Fish and Wildlife Service makes recommendations and advises the regulatory agencies on ways to avoid or minimize the effects of impoundments on fish and wildlife. The Endangered Species Act prohibits actions that would harm endangered species; it can be directly enforced by the U.S. Fish and Wildlife Service (16 U.S.C. §1538).

SUMMARY

Standards for impoundments and refuse piles were developed after the Buffalo Creek disaster in February 1972. The Federal Mine Safety and Health Act establishes mandatory standards to protect the health and safety of miners. This limitation is important in assessing MSHA's regulatory responsibility and authority associated with potential impoundment breakthroughs and failures that do not affect the health and safety of miners but may affect persons, communities, or ecosystems downstream. On the other hand, the legislative and regulatory direction of SMCRA, to protect society and the environment from the adverse effects of surface coal mining operations, complements the intent of the Federal Mine Safety and Health Act.

The material reviewed in this chapter indicates that a substantial regulatory structure covers many aspects of the design, construction, and operation of coal waste impoundments. *The committee concludes that while the regulatory review of a proposed impoundment is detailed with respect to the embankment, the regulatory review of the impoundment basin has been less rigorous.* The authority for review of the basin characterization and design appears to be covered only in general language authorizing investigation of all relevant issues with respect to the impoundment. **The committee recommends that MSHA and OSM should have clear authority to review basin design.** It is not evident to the committee whether specific legislation to authorize more detailed examination of basin issues is required, or whether these issues can be handled by additional rulemaking under existing authority.

The committee heard reports that the review process is undesirably lengthy, commonly exceeding 2 years. In addition, it appears that review times have lengthened considerably in recent years. *The committee concludes that timely review is an essential component of an effective regulatory process.* An efficient and coordinated regulatory review process can be of substantial benefit to both the applicant and the jurisdictional agencies. A well-coordinated technical review process can ensure that the health and safety of both the miners and the public, and the protection of the environment are ensured in a sensible and streamlined way. **Therefore, the committee recommends that the review process for both new permits and revisions to existing permits be overhauled to include the following elements:**

- **A formal joint process that would coordinate the currently fragmented and inefficient collection of reviews into a single process.**
- **Sufficient staff for engineering and other reviews in the agencies that participate in the joint process so that the time required to complete the review can be reduced significantly.**

3

Planning Coal Slurry Refuse Impoundments

This chapter summarizes the current engineering elements of the design, operation, and reclamation of coal refuse impoundment systems. The chapter addresses these elements in relation to the Appalachian coal region, where impoundments are often located in steep valleys. It is not a comprehensive overview of all engineering design options, but a description of the major elements currently implemented in the planning and design of coal slurry refuse impoundments. The engineering design process is specific to the conditions of each particular site: What works in one circumstance may not work in another.

The position of potential fluid pathways—such as coal seams, mine workings, and fractures—relative to an impoundment is a significant factor in the design of new and modifications to existing coal refuse impoundment systems in the Appalachian region. Most impoundments in Appalachia utilize the natural topography to form the storage basin that will contain the slurry. This is often accomplished by constructing an embankment at the mouth of a small valley or watershed to complete the basin structure used for storage. The stream channel at the base of the basin selected for the impoundment defines the approximate level of local drainage. Coal seams and mines that do not crop out above the level of the stream channel are termed below-drainage, whereas, those that crop out along the valley wall above the stream channel are termed above-drainage (Figure 3.1). In some locations, both above- and below-drainage coal seams (and mines) may be present.

The relative elevations of local drainage and slurry height can be critical. For example, during the filling of the impoundment, if the slurry elevation exceeds the level of an above-drainage coal mine, the mine could become submerged and provide a hydraulic conduit for the slurry. In addition, below-drainage workings can be in hydraulic connection with slurry through less direct means, such as fractures. Both above- and below-drainage mine workings require specific engineering considerations.

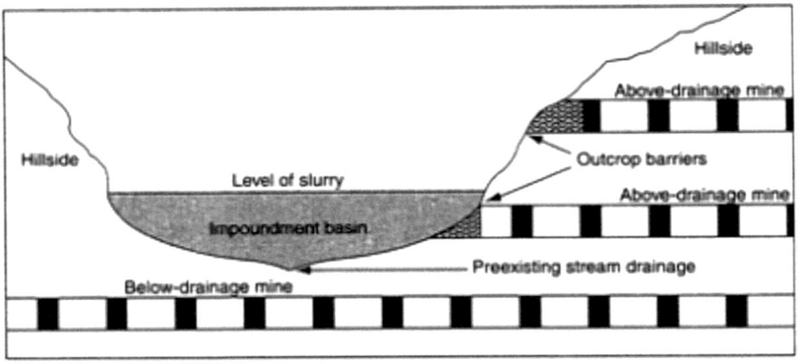

FIGURE 3.1 Cross section of schematic coal waste impoundment, depicting underground mine workings above and below drainage with respect to the impoundment; outcrop barriers are indicated.

For impoundments with both above- and below-drainage mine workings, assessment of the strength and permeability of the material between the impoundment and the underlying mines is important. This assessment includes the distance and geologic conditions between the impoundment and the underlying mines, the potential for hydraulic connection, and the potential for collapse of the underground workings. At sites where above-drainage seams have already been covered or inundated by slurry, this assessment is more complex. Finally, for impoundments that will cross above-drainage coal seams or workings in the future, it is essential to explore thoroughly the site along the coal seam outcrop and to plug or seal off the contacts with the coal seam (see Chapters 5 and 6).

Proper engineering is critical at all stages of impoundment life: site assessment, construction, operation, monitoring, and closure. The sections that follow describe current approaches to each of these topics.

GENERAL IMPOUNDMENT SITING CRITERIA

Site investigations require a preliminary examination for site selection followed by a detailed site study to develop safe and economical designs that satisfy regulatory requirements. These investigations are primarily focused on siting the embankment by assessing the factors influencing embankment foundation strength, water seepage through the embankment foundation and the bottom of the impoundment, and the quality and quantity of foundation materials. Specifically, these factors include the geology and hydrogeology of the site, the potential for subsidence, and the geotechnical characteristics of foundation materials.

While the geotechnical characterization of the embankment foundation is well developed in the coal industry, the level of geotechnical investigation of the basin seems to exhibit much less consistency. Basin investigation can evaluate potential weaknesses along which fluids could travel. Examples include rock fractures, subsidence zones, and permeable strata. Although not common, some coal-bearing formations contain limestone, and there is the potential for karst development and fluid migration along solutionally-enlarged fractures or conduits. The committee could not identify a consistent set of criteria or a guidance document for conducting a site investigation of the impoundment basin. Regulations to this effect are usually promulgated by the states and vary by state. However, despite this lack of regulation, standard engineering practice implies that some level of basin analysis should be included in design; the degree to which this is done appears to be site specific.

The probability of exposing cracks, faults, coal seam outcrops, or other preferential pathways for water is greatly enhanced by clearing surface soils and exposing the basin foundation materials. Because the topography of the Appalachian coal region is characterized by steep surfaces covered with dense vegetation, the side slopes of the basin area are often difficult to access using conventional earth-moving equipment. As such, many basin investigations appear to be limited to visual reconnaissance of the basin side slopes, with clearing of vegetation and topsoil and stripping of soils limited to the accessible areas in the valley bottoms only.

Site Geology and Hydrogeology

Geologic site investigations should include the depth, thickness, continuity, and composition of each significant geologic layer and an evaluation of regional and local fractures, faults, and lineaments. Rock and fracture fluid pathways can become significant zones of weakness for both the impoundment basin and the embankment. Basic regional geologic information in the form of maps and stratigraphic data (e.g., coal seam identification) can usually be obtained from the state geological survey. The principal methods used to determine detailed bedrock and soil stratigraphy are drilling and sampling. Geophysical methods, such as seismic refraction and electrical resistivity, may complement drilling and sampling (see Chapter 5).

These methods can prove invaluable for determining bedrock surface and water table profile. However, they require careful interpretation in conjunction with other geologic information, particularly that obtained from boreholes.

In the Appalachian Plateau, the underlying bedrock is strongly fractured, and nearly all of the rock types found in coal basins have very low primary porosity (Stach, 1982; Weinheimer, 1983). Continuous regional joints and fracture systems are superimposed on the relatively shallow near-surface fracture systems, resulting in a complex array of fractures that dominate the shallow flow of groundwater (Minns, 1993). The majority of water in coal-bearing rocks is transmitted in secondary permeability features, such as fractures, joints, bedding planes, and coal cleats. The thin rind of fractured bedrock that extends from the bedrock surface down to a depth of between 80 and 200 feet, has been termed the near-surface fracture zone (Kipp et al., 1983; Wyrick and Borchers, 1981). The role of secondary permeability decreases with depth due to decreasing fracture width, length, and interconnectedness. Therefore, permeability of the original rock gradually becomes more important with depth (Harlow and LeCain, 1991). Hydrogeologic studies of boreholes in coal-bearing strata in the Appalachian coalfields have consistently shown that fractured rock and coal seams have permeabilities nearly three orders of magnitude greater than that of adjacent, nonfractured bedrock (Harlow and LeCain, 1991; Kipp and Dinger, 1987; Minns; 1993; Wunsch, 1993).

Subsidence

Mine subsidence is defined as the ground movement that occurs when the overlying strata collapse into the mine openings (Brauner, 1973; Gray et al., 1974; Shadbolt, 1977; Singh, 1992). These can be either active or abandoned mines. These collapses can create zones of weakness or fluid flow, which are important considerations in impoundment design. On the surface the disturbance is manifested as cracks depending on the depth-to-mined-height ratio and other factors. The layers above the caved-in zone are subjected to compressive and tensile stresses (Figure 3.2). Tensile stress in the vertical direction generally gives rise to bed separation, whereas, in the horizontal direction it may open joints in the rock formations. Surface subsidence entails both vertical and lateral movements of four types: cracks (tension or shear), buckling (due to compression or shear), pits (also locally termed potholes, sinkholes, chimneys, crownholes, or pipes), and troughs (or sags).

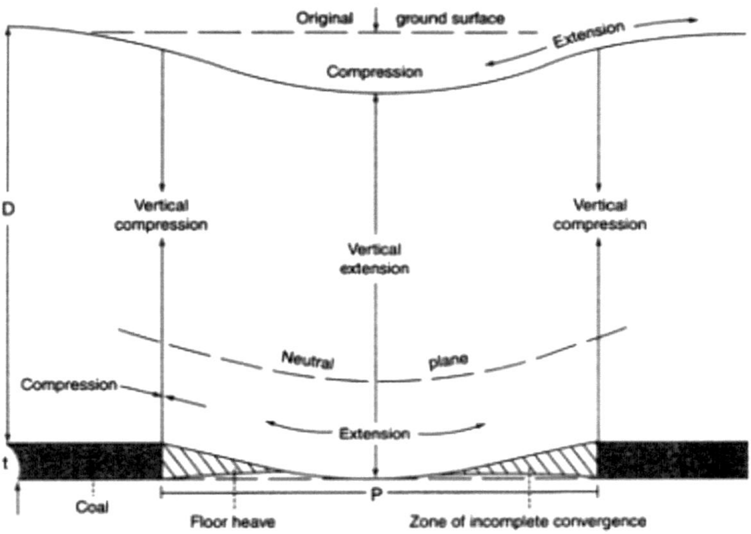

FIGURE 3.2 States of stress over a mined area. A generic diagram that encompasses active and abandoned mines, and longwall and room-and-pillar mines. D=depth to seam; t=mined coal thickness; p=mined area (longwall or pillared) within coal seam. After Shadbolt, 1977.

The magnitude and extent to which subsidence will impact an area are governed by a variety of factors. For example, coal seam geometry, geology and hydrogeology, and mining history may directly affect the potential degree of subsidence from coal mining. The amount of coal removed from the seam, the depth of the mine, and the orientation of the void, respectively, determine the degree, timing, and geometry of subsidence features. The mining methods used, rate of face advances, degree of extraction, and presence or absence of mined areas can also have an effect. Finally, factors such as in-situ stresses, topography, and time elapsed since the area was mined can also influence subsidence.

Site-specific geologic information, including the availability of mine maps and associated geotechnical information, is critical to evaluating the potential for subsidence. These characteristics apply to both active and abandoned mines. An understanding of these phenomena is also required for planning mine layouts near or under previously existing impoundments.

Coal Barriers

Another potential ground weakness resulting from mining may occur near above-drainage workings. Coal pillars are sized to protect mine openings from overburden stresses. Individual pillars protect the adjacent openings, and large "barrier" pillars are left in place to protect groups of entries or to assume abutment loads from pillar mining. Boundary or perimeter pillars are left in place to isolate the mine workings from the effects of stress associated with adjacent mines, and from the effects of water or gases in adjacent mines. In mines that extend close to the outcrop, a boundary—perimeter barrier or outcrop coal barrier—is left in place to contain any water that collects in the mine in the future. Similarly, when mining close to the outcrop, a solid coal barrier is left in place to contain any such water. If the outcrop barrier width is insufficient, a blow-out or outcrop barrier failure can occur. A blow-out failure occurs when water pressure within the mine forces the outcrop barrier outward, or when deterioration of the floor enables the outcrop barrier to slide. Blow-ins could occur where the pressure of water or slurry impounded above the outcrop barrier forces the barrier into the mine. A breakthrough is a general term that describes the catastrophic failure of outcrop barriers, resulting from water or slurry entering the mine through fractures, joints, "mud seams," or lineaments in the overburden strata. If the mine is active, a breakthrough could endanger miners. Water or slurry in a mine creates the potential for stream pollution and consequent endangerment of public safety.

Currently, no federal (but some state) regulations govern the width of outcrop barrier that should be left. For example, Kentucky guidelines require a barrier of 50 feet plus 1 foot for each foot of hydraulic head; West Virginia does not specify a minimum but requires mine maps and permits to note the outcrop barrier width (Table 3.1).

Geotechnical Characterization of Foundation and Construction Materials

Assessment of the geotechnical properties of soil and rock materials forming the foundation for both the embankment and the impoundment basin is a critical part of the site characterization and design process (Vick, 1990). Embankment foundations receive special consideration in MSHA reviews since they must support loading, control seepage, and satisfy structural and water management issues. In addition, seepage from the impoundment often is controlled by the permeability of the underlying natural rock or soil strata.

TABLE 3.1 Coal Outcrop Barrier Minimum Width, by Selected States

State	Minimum Width in Feet	Regulations
AL	No standard minimum width is specified.	Alabama Administrative Code, Ch. 880-X-81
IL	No standard minimum width is specified.	Illinois Administrative Code, Title 62
IN	No standard minimum width is specified.	Indiana Code, Title 14
KY	Minimum width is to be determined on a case by case basis but will be at least W=50 feet + H, unless it can be shown that less will be effective. An exception may be made if no water accumulation is expected.	Kentucky Administrative Regulations, Title 405, Subtitle 18:010.6
MD	Minimum width is to be 50 feet or W=20 + 4T+0.1D (whichever is greater). If barrier will be affected by hydrostatic head, the barrier will be a foot wide for each foot to hydrostatic head.	Code of Maryland, Title 26, Subtitle 20.13.11
OH	No standard minimum width is specified.	Ohio Administrative Code, 1501:13
PA	No standard minimum width is specified, but the necessity for a coal barrier to prevent water release is noted.	Pennsylvania Code, Title 25, Chapter 89.54 (b)
TN*	No standard minimum width is specified.	30 CFR, Part 942
VA	Minimum width will be determined either as a site specific design or by W=50+H. An exception may be made if the barrier will not be subject to hydrostatic head.	Virginia Administrative Code, Title 4, Agency 25, Chapter 130–817.41 (a)(i)(3)
WV	No standard minimum width is specified; however, the plans must note the outcrop barrier width and the maximum head. In addition, mine maps must designate the location and thickness of outcrop barriers.	West Virginia Surface Mining Reclamation Regulations, Title 38-Series 2-Section 3.13 (a) and (c) and Section 3.4 (f.2.H).

NOTE: W=minimum width (feet); H=maximum hydrostatic head (feet); T=thickness of extraction (feet); D=depth of overburden (feet)

* Tennessee does not have primacy.

The geotechnical properties of embankment foundation materials are determined by physical tests to assess the compressive and shear strengths, bearing capacity, permeability, consolidation and cohesion, plasticity, and moisture content of the potential foundation materials. In particular, determining the permeability of the surficial materials that underlie the impoundment is necessary for assessment of possible seepage from the pond; this is especially important if a mine underlies the impoundment.

Geotechnical investigations also involve drilling in the "footprint" of the embankment. Drilling programs are designed to penetrate the foundation materials to an appropriate depth. Multiple holes are required along the embankment footprint to establish homogeneity or variability of the foundation materials throughout the structure. Materials recovered from the drill hole are tested to establish strength, consolidation, and permeability characteristics.

Settlement of the embankment foundation is generally not a factor in embankment design where the foundation is sound rock and the embankment area has not been undermined. Although many embankments are constructed from noncohesive materials and are not subject to cracking, some compacted refuse materials exhibit brittle behavior and may crack as the foundation material settles. Particular attention is given to the foundation of both starter dams and embankments to avoid embankment cracking, which can be caused by differential settlement between a compressible valley bottom and rock abutments or outcrops.

In some cases, the geotechnical properties of the basin foundation materials are also characterized. The basin area is not always drilled, but where subsidence of below-drainage mine workings is a possibility, operators sometimes drill and test as they do for the embankment foundation. Boreholes in the basin area must be plugged to limit communication between the impoundment and the underlying strata.

Finally, geotechnical characterization of the construction materials to be used for the embankment is also a routine part of site evaluation. Most coal operators utilize the coarse refuse generated by the coal preparation process to construct the embankment. Using coarse refuse for embankment construction solves the disposal issue for the coarse waste, which would otherwise have to be disposed of elsewhere within the mine's permitted area. Coarse refuse is fairly homogeneous in particle size and strength characteristics over time and is therefore a comparatively predictable construction material for meeting the engineering design specifications for embankment materials.

IMPOUNDMENT DESIGN AND CONSTRUCTION

Planning for a coal slurry impoundment includes the design of two major elements, the embankment and the basin. Information gathered from the site investigation plays an integral part in this process. The regulations MSHA established require detailed investigation of the stability of the embankment but are less specific concerning the basin structure. In the subsections that follow, typical design procedures for each are described.

Embankment

MSHA classifies three types of impoundments: cross-valley, diked, and incised (Figure 3.3). For coal slurry impoundments in the Appalachian coal region, the cross-valley impoundment is predominant (ICOLD, 1996; MAC, 1998; U.S. Army Corps of Engineers, 1994; Vick, 1990). The cross-valley impoundment consists of an embankment constructed across a valley, with slurry discharged within the valley upstream of the embankment. The operation of a refuse impoundment normally includes staged construction of the embankment, generally by upstream or downstream methods.

The oldest and most commonly used method of embankment construction is the upstream method (Figure 3.4), where the embankment centerline is moved upstream with sequential raises. This method is used if suitable materials are available to build a starter dam, if the seismic hazard is low, and if the embankment can be raised in a stable manner. With this method, a starter dam is constructed using coarse refuse or locally available materials, and fine refuse is discharged hydraulically from the crest of the starter dam to form a beach. Coarse refuse is pushed out over the beach area of the impoundment and is compacted to form the foundation for a second embankment raise. Construction continues in this manner as the embankment increases in height. One disadvantage of this type of embankment construction is that the sections of the embankment constructed at later stages lie above finer-grained material discharged during the preceding stages. Under static loading conditions, the ultimate embankment height will depend on the strength of the consolidated fine refuse within the zone of shearing, the steepness of the downstream slope of the embankment, and the location of the phreatic surface within the embankment. Under seismic loading, the stability of the embankment depends on the potential of the consolidated fine refuse to liquefy. Upstream construction lends itself to concurrent reclamation on the downstream face of the embankment, because the downstream face is usually designed at the final reclaimed configuration of the embankment.

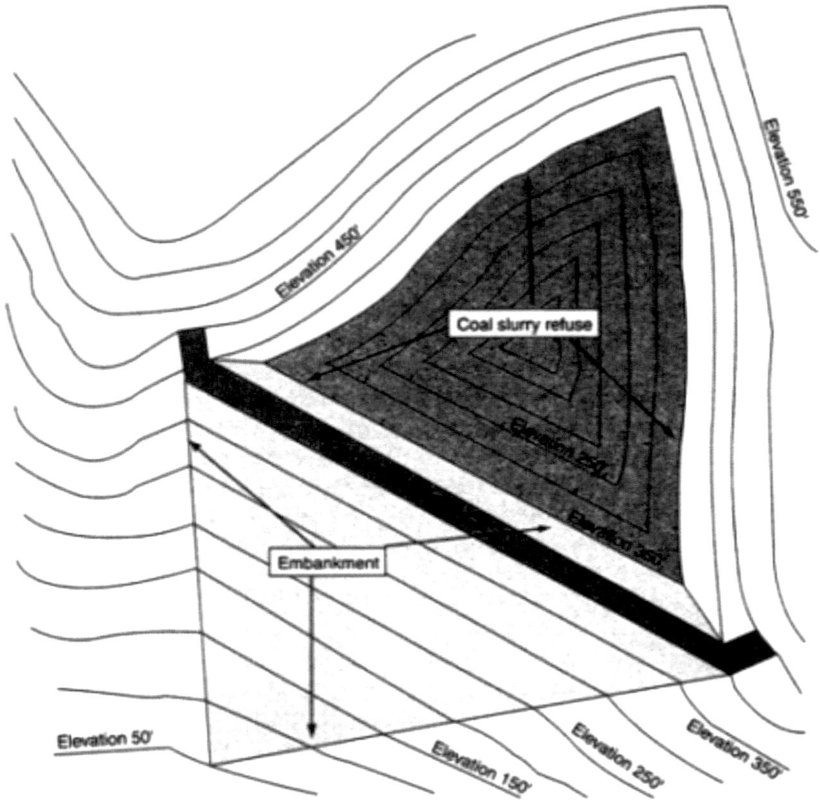

(a) Cross-valley refuse impoundment.

FIGURE 3.3 Schematic diagram of coal refuse impoundments: (a) cross-valley, (b) diked, and (c) incised impoundments.

(b) Diked refuse impoundment.

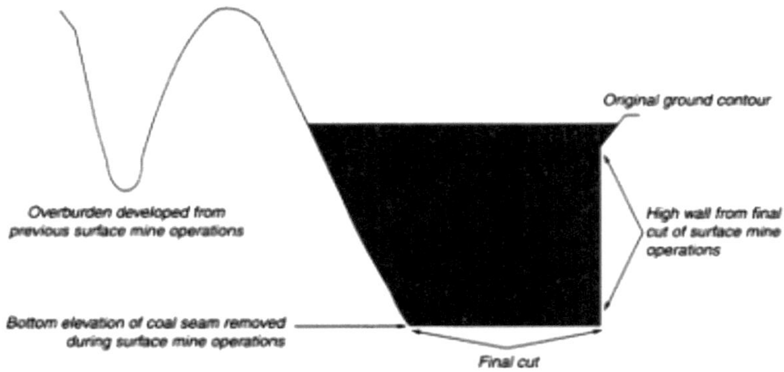

(c) Incised refuse impoundment.

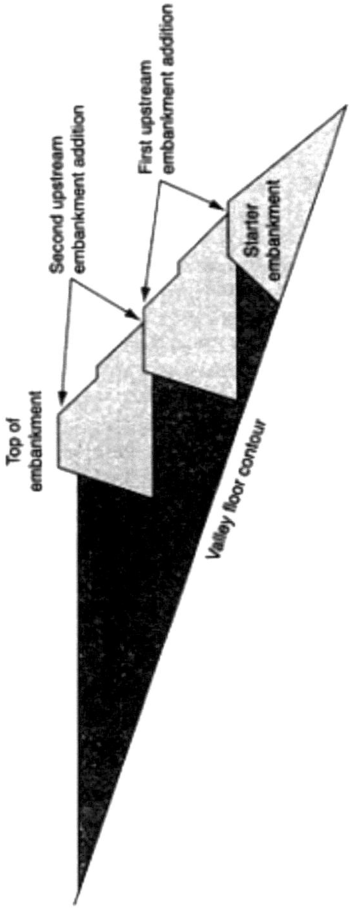

FIGURE 3.4 Sequential raising of a tailings dam by the upstream method.

The downstream method consists of construction such that the centerline of the embankment is moved downstream with subsequent raises (Figure 3.5). For this method, the embankment is not underlain by previously discharged fine refuse, and it is generally more stable than an upstream embankment. However, downstream construction requires increasing volumes of embankment material with subsequent embankment raises. Sedimentation ponds, pump stations, and other facilities downstream of the embankment may conflict with future structural fill using downstream construction methods. A drawback of downstream construction is that the embankment cannot be reclaimed until closure, since the embankment face is continually being added to as the structure moves downstream. For these reasons, the downstream method is less attractive than the upstream method for construction of coal slurry refuse impoundments.

The centerline and modified centerline methods of embankment construction are a compromise between the upstream and downstream methods, in that the crest of the embankment is raised vertically, or nearly vertically, instead of being displaced upstream or downstream (Figure 3.6). This method is generally between the upstream and downstream methods in terms of stability and required volume of embankment material.

Embankment stability in coal slurry waste impoundments is evaluated with the same techniques and criteria used for water-storage dams (U.S. Army Corps of Engineers, 1982; Wilson and Marsal, 1979). Much like a water-storage dam or reservoir, the embankment of a coal slurry impoundment retains a saturated material having low shear strength (D'Appolonia Consulting Engineers, 1975). MSHA regulations are based on the stability of the embankment, foundation, and abutment; control of seepage for hydraulic considerations; and management of excess water to control embankment overtopping.

Coal refuse impoundments are subjected to embankment stability evaluations under static conditions for the designed construction, operation, and closure conditions. The evaluation of embankment slope stability results in a calculated factor of safety, which MSHA requires the impoundment designer to provide, along with methods used to obtain. The factor of safety is typically 1.5, but is depends on site conditions (30 CFR 77 § 216–2 (a)(13), similar to those used for water-storage dams.

Evaluation of seismic stability of the embankment (ICOLD, 1989a; Seed, 1979) is based on the potential seismicity of the site and the expected response of the embankment to seismic vibration. Seismic activity can also affect the stability of pillars and the strata overlying the coal seam. While seismic hazards are generally low in Appalachia, they are not absent (Frankel et al., 1996). Mining-related activities such as underground blasting or

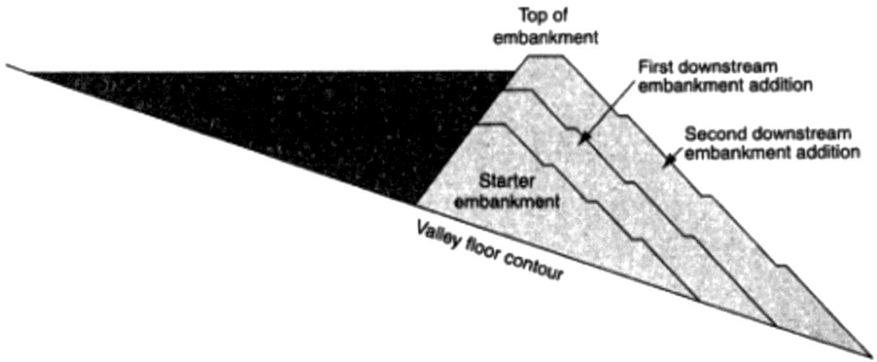

FIGURE 3.5 Schematic diagram of sequential raising of a tailings dam by the downstream method.

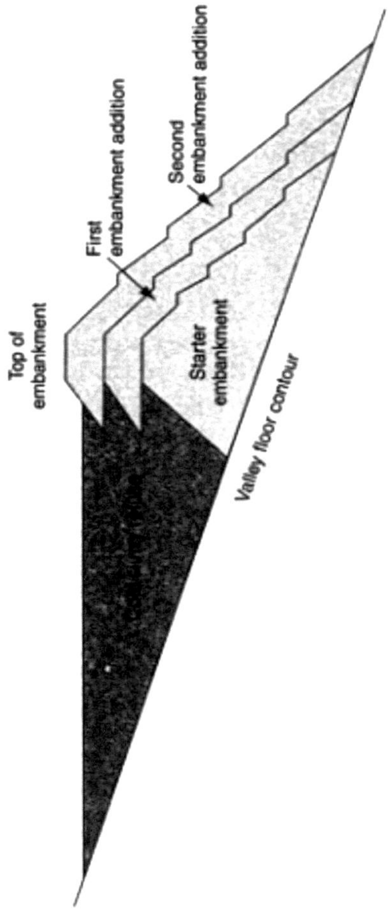

FIGURE 3.6 Sequential raising of a tailings dam by the centerline method.

unplanned explosions or pillar failures can produce vibrations, but are generally of smaller magnitude and shorter duration than actual seismic events. The duration of the seismic motion and the accelerations over this duration influence the potential for liquefaction, displacement, and stability of the refuse in the impoundment.

Another key factor in the static and seismic slope stability of the embankment is control of seepage for hydraulic considerations (subsurface water pressure, piping, and erosion), as well as downstream water quality aspects. The embankment is designed to control water pressure on the upstream side of the embankment and to prevent migration of fines and to minimize water pressure and potential for piping on the downstream side of the embankment. This can be done through filter and drain zones within the embankment that collect and route water to the downstream toe of the embankment. While basal drainage beneath the embankment is common in upstream construction in Appalachia, internal filter and drain zones are not.

Management of excess water (by embankment freeboard, upstream diversion, and controlled water removal) can prevent embankment overtopping. Overtopping of an embankment occurs when the water inflow volume exceeds the storage capacity of the impoundment or spillway. Overtopping is prevented by providing sufficient freeboard or allowing for a sufficient spillway discharge rate to either store or safely discharge the anticipated inflow to the impoundment. To increase the stability of an embankment, inflow can be reduced with a diversion channel to direct upstream runoff past the impoundment.

Basin

Filling the basin with slurry increases stress on the basin floor and walls and increases hydraulic pressure. In response to this increased stress, compressible materials, such as unconsolidated surface soils, settle. In areas overlying underground workings, the increase in stress may cause significant differential settlement near the workings themselves or a zone of subsidence above the workings.

The impoundment system's hydraulic performance is generally based on the comparative permeabilities of the coarse refuse embankment, the stored fine slurry, and the foundation materials beneath the entire impoundment. In the Appalachian coal industry, the drainage of the refuse slurry water and the pore water entrained in the fine refuse is primarily by upward water migration as the slurry settles and consolidates. The collected water then evaporates or is decanted for reuse in the preparation plant. Drilling into fine refuse materials in impoundment basins has revealed that the strength characteristics of the stored solids tend to increase with depth because the pore water entrained in fine refuse decreases with depth (Schiffman et al., 1988; Thacker, 2000; Thacker et al., 1988). Common practice relies on the self-sealing of the basin with slurry compaction and consolidation rather than the construction of liners.

SLURRY AND WATER MANAGEMENT

In most refuse impoundment systems, the slurry is discharged into the basin area through a pipeline system along the embankment crest. Slurry is either released into the basin from a single-point discharge that is moved periodically across the embankment face, or it is discharged from multiple points along the embankment. Both methods of discharge create a beach effect, with coarser material close to the discharge point, and finer slurry and water usually farther from the discharge point at the upstream end of the impoundment. The water is removed from this area through a decanting or pumping system.

The consolidated beach that forms against the embankment will enhance its stability because the coarser material tends to drain more easily and is the more competent component of the slurry. Most coal operators limit the location of slurry discharge to the embankment face so as to improve the stability of the embankment. However, where geologic anomalies or coal seam outcrops occur, advantages can be gained by managing slurry deposition around the basin perimeter. Depositing slurry so that it forms a consolidated beach around the basin perimeter creates a control zone, which forces the separated water away from the basin foundation contact. This reduces the potential for hydraulic communication between the fully saturated pool area and an exposed coal seam, crack, or opening in the basin area.

Operation of a coal waste slurry impoundment requires management of the slurry liquid without discharge into the surrounding environment, as well as accommodation of natural precipitation and evaporation. Effective water management for mining facilities (Hutchinson and Ellison, 1992; Vick, 1990) includes water balance, water reclamation, and seepage and underdrainage control.

Water balance is determined by measuring the inflow and outflow of the impoundment system on an annual to daily basis. The water balance can be used to evaluate whether there is a long-term net gain or loss of water under anticipated climatic conditions and to assess the effects of particular short-term extreme precipitation events The balance is also used to ensure that the amount of water reclaimed is maximized for reuse in the preparation plant (or other beneficial use).

Collection of reclaimed water consists of returning water from the impoundment back to the preparation plant or other use. Where possible, this process is optimized to promote settling of suspended solids and to minimize the amount of suspended solids in the reclaimed water.

Finally, control of seepage and underdrainage minimizes the amount of water that leaves the impoundment through the foundation. The objective is to collect and reuse the maximum amount of water.

IMPOUNDMENT SYSTEM MONITORING

Impoundment monitoring, which continues until closure, includes both required inspections and the utilization of instrumentation to detect changes within the system. After closure, both the instrumentation required and the frequency of monitoring are reduced. Impoundment inspection usually includes a visual examination of the entire impoundment area, including the embankment, basin, and proximate surroundings. The embankment inspection includes looking for cracks, seeps, slumping, or other unusual conditions. Monitoring instrumentation is often used to measure water pressure within the embankment (ASCE, 1999; Dunnicliff, 1988; Penman et al., 1999; U.S. Army Corps of Engineers, 1995; U.S. Bureau of Reclamation, 1987; USCOLD, 1993). The water level in the basin pool is measured and instrumentation data are reviewed to assist in water management. In addition, other monitoring equipment can be installed to detect movement within the embankment or subsurface. Measurements are commonly made to detect the following effects:

- Surface Displacement—Measured most often by conventional surveying equipment to detect vertical and horizontal displacement.
- Internal Movement—Measured by single- and multi-point extensometers, continuous profile gauges, inclinometers, tilt-meters, transverse-acting devices, and time domain reflectometers (to determine where beds separate) to detect vertical and horizontal movement below the surface of the impoundment slurry level.
- Pore Pressure—Measured by piezometers—including open, well-point, and closed types—to detect the pressure exerted on the instrument. This pressure is then balanced against an equivalent hydrostatic head or equivalent fluid pressure.

- Groundwater Levels—Measured by monitoring wells. In addition, the groundwater chemistry may also be monitored.
- Surface Water Discharge—Measured by weirs, flumes, or flow meters.
- Subsurface Settlement, Readjustment, and Subsidence—Measured by borehole extensometers, time domain reflectrometers, accelerometer arrays, and piezometers to detect changes in hydrostatic pressure. Visual observations are also used to identify subsidence by propagation of fractures to the ground's surface.

CLOSURE AND RECLAMATION

The process of changing from an active or inactive impoundment to an abandoned impoundment is referred to as closure and reclamation (Sweigard, 1992). Three major elements comprise reclamation: surface grading, closure water management, and long-term stability. Surface grading reconfigures the final impoundment surface so it sheds runoff and will not erode or form pools of water. This may require major regrading of solids or selective slurry discharge during the final stages of operation to create the desired draining surface. Cover materials such as coarse refuse or soils from surrounding locations are placed over the consolidated mine refuse mass and graded to the final closure configuration. The regraded surface is then covered with topsoil or an approved substitute material and revegetated.

Closely tied with surface regrading is closure water management. This involves: removal of the residual ponded slurry water (by reuse or evaporation) before surface regrading; control of adverse geochemical reactions, sometimes using chemical or other additives, and capping; and management of runoff to control erosion and sediment before and during establishment of vegetation. Long-term stability is maintained as the impoundment is transformed from an operating facility with active observation and monitoring, to a closed facility with less frequent observation and monitoring. This means that the slope stability and erosional stability of the closed and reclaimed facilities must not deteriorate over time. The reclamation surety for slurry impoundments is typically released in phases. As an operator completes a discrete phase of the reclamation obligation, such as earthmoving or revegetation, portions of the surety are released. When the final phase is released, the company has no further liability for the reclaimed impoundment, unless the regulatory authority can show that the operator misrepresented submitted material facts (L.Adams, Kentucky Department of Surface Mining, Reclamation, and Enforcement, personal communication, 2001).

SUMMARY

The principles that govern regulation of the design of structures that were promulgated in response to the Buffalo Creek disaster are well understood and fully documented. *While continued vigilance concerning design, construction, and operation of impoundments is clearly warranted (Chapter 6), the committee concludes that uncertainties remain in the characterization of the basin area and, therefore, in the mitigation of risks associated with the breakthrough potential.* The potential for underground coal mine workings to be in close proximity to an impoundment is a factor in the design of new and in modifications to existing coal waste impoundments in Appalachia. The relative elevation of local drainage and slurry height, with respect to underground mines, can be critical. Existing impoundments with above-drainage mine workings, where the outcrop slurry elevation does not exceed the level of the coal mine workings, can incorporate mitigating measures for these workings in their design relatively easily (see Chapter 6). Above-drainage coal mine workings in existing impoundments, where the slurry elevation exceeds the level of the mine workings are the most challenging in the design and operation of a facility.

4

Mine Mapping and Surveying

A key element of assessment of the potential for breakthrough of coal slurry materials into underground mine workings is the accurate delineation of the extent of those workings with respect to the ground surface in the impoundment's basin area. This chapter deals with several aspects of that problem. The adequacy of existing mine maps and recommendations for the storage and preservation of maps make up the first part of the chapter. While mine maps provide information critical to the characterization of a site, there are significant limitations to some maps, particularly those for abandoned mines and mines operating before 1969. In addition, standard practice in mapping and surveying varies from site to site. When questions concerning the accuracy of maps remain, additional effort to locate underground workings is warranted. Chapter 5 reviews geophysical methods that can be applied to locate underground mines.

SURFACE MAPS

There are two primary sources of surface topographic mapping are the U.S. Geologic Survey's 7 1/2-minute quadrangle maps and aerial topography. The U.S. Geological Survey's contour intervals vary according to the steepness of the terrain. A 40-foot contour interval is common for southern Appalachia. The generally accepted accuracy in surface elevation using these maps ranges from 20 to 40 feet. In contrast, mine maps are typically maintained to a tenth or hundredth of a foot.

However, coal companies commonly use aerial photography to determine topography for critical construction projects such as refuse impoundments, preparation plants, and mine portals. The contour intervals for aerial topographic measurements usually vary in sensitivity relative to the steepness of the local topography. It is common to have one or two foot contours for relatively flat terrain and as much as 5 to 10 foot in mountainous areas. Contours are generated from surveyed control points flagged at the time of the flight. Maps so determined also show man-made features such as roads, excavations, and buildings.

Where closed-loop underground mine maps are available (see below), the accuracy in depicting the surface topography controls the accuracy by which the coal outcrop is located on a hillside. **The committee recommends that adjacent to existing or proposed refuse impoundments, the coal outcrop location(s) be determined using aerial topographic measurements.**

UNDERGROUND MINE MAPPING

The use of inaccurate or incomplete mine surveys and maps may result in construction of an impoundment in an area not known to have been mined; if unknown mine workings are present, the impoundment could suffer unexpected structural failure (Franklin et al., 1997). In areas where impoundments are constructed near known or suspected underground mines, vertical and horizontal barrier distances between the mined area and the impoundment, as depicted on old maps or surveys, may not be accurate. Regulations promulgated as a result of the Federal Coal Mine Health and Safety Act (30 C.F.R. § 75.1200–2 (b)), require closed-loop mine surveys. However, surveys for many older mines were not closed. Furthermore, underground mine surveys may have been based on a foreman's notes or sketches that lack a reference point or a recognized coordinate system and therefore cannot be accurately located. These shortcomings are more common in small mines or room-and-pillar mines where a number of short panels were driven and extracted. Compounding the problem, some maps and records of older mines have been lost or destroyed.

The condition and thickness of a barrier, whether composed of coal or other rock type, between an underground coal mine and a surface impoundment can be difficult to determine in the steep topography common in Appalachia. Yet, this information is needed to quantify the potential for breakthrough. The tools available to determine the extent of mining in an abandoned mine and the condition and thickness of the outcrop are not highly advanced. Although extensive drilling can be used, it may miss smaller mined or disturbed areas. In contrast to invasive drilling, geophysical techniques can obtain useful information without penetrating the Earth's surface. The objective of geophysical surveys is to constrain the physical characteristics of some three-dimensional volume of earth material, including the presence of voids. Although no geophysical technique performs optimally under all geological and topographic conditions, multiple geophysical techniques may be necessary to reduce the probability for error to an acceptable level (see Chapter 5).

The primary use of underground mine maps is to determine accurately the dimensions of pillars and mine openings and to locate the mine with respect to the surface and any horizontally or vertically adjacent surface or other underground mines (Shackleford, 2001). The following information relevant to site characterization is included on a typical underground mine map (30 C.F.R. §§ 75.1200, 1200–1):

- All pillared, worked out, and abandoned areas, pillar locations, sealed areas, future projections, adjacent mine workings within 1,000 feet, surface or auger mines, mined areas of the coalbed, and the extent of pooled water;
- Dates of mining, coal seam sections, and survey data and markers;
- Surface features (e.g., railroad tracks, public roads), coal outcrop, and 100-foot overburden contour or other prescribed mining limit;
- Mineral lease boundaries, surface property or mine boundary lines, and identification of coal ownership.

A coal section, which describes the mined thickness of the sequence of coal, rock, and partings, is listed on the mine map. Coal sections are typically recorded when a survey spad is set (see definition in glossary). The coal section permits calculation of mined coal tonnage, percentage of coal recovery, and percentage of reject (in-seam and out-of-seam rock). An example of a portion of a typical underground mine map is shown in Figure 4.1.

Underground and surface mine maps are collected and stored both on the state level and by MSHA and OSM. Operators of underground mines are required to submit maps to MSHA at least annually for the approval of ventilation plans. These maps are maintained at the various MSHA district offices or archived at a central location until the mine closes. Following mine closure, a copy of the final map is forwarded to the OSM National Mine Map Repository in Pittsburgh, Pennsylvania (see below). Therefore, MSHA is a source of maps for active mines only.

There is considerable activity at the state level concerning mine maps. For example, the Commonwealth of Virginia has embarked upon an ambitious program to accumulate mine maps and place them in a digital database (Sidebar 4.1). Related activities underway in West Virginia and Kentucky are discussed in Sidebars 4.2 and 4.3. Comparison of the mapping

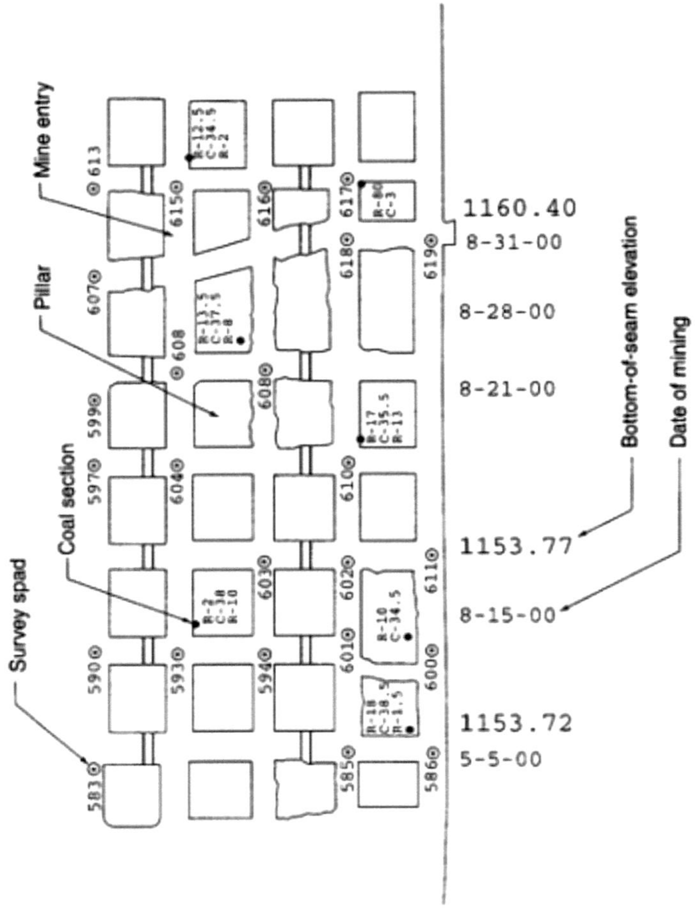

FIGURE 4.1 Typical mine map showing a schematic plan view of underground mine workings. (R=rock, C=coal).

SIDEBAR 4.1 MINE MAPS AND STORAGE, COMMONWEALTH OF VIRGINIA

Virginia's mine map repository is coordinated through the Virginia Department of Mines, Minerals, and Energy and relies upon the Division of Mineral Resources and the Division of Mined Land Reclamation. It contains 4,226 digitized and geo-referenced mine maps, representing 70 to 80 percent of the mined-out areas in the state. The remaining mines are not yet included because the maps lack sufficient information for accurate location, because mining occurred before Virginia required map records, or because Virginia has not yet acquired the maps from coal companies.

The project began with 10,000 folded blueline mine maps and 30,000 microform or microfilm records. These maps ranged from the product of trained surveyors, referenced to state plane coordinates or local coordinates, to hand-drawn sketches lacking scale, a coordinate system, and reference points. Eliminating duplicates and earlier versions of the same mine was the department's initial task. Where possible, the Virginia Department of Mines, Minerals, and Energy locates mines using a global positioning system. In an effort to characterize errors in mine location, the department conducts random checks using the global positioning system and coal company survey records. The error for these maps is a function of the surveying accuracy, which in turn is related to the age of the mine. The closure error of mines using modern surveying techniques is 1:5,000 to 1:15,000. While older mines may not have been surveyed or if surveyed, a closed loop may not have been used to quantify error. Accuracy is predicated upon the methods used to locate the mine. The Virginia Department of Mines, Minerals, and Energy estimates the following degrees of accuracy:

- ±500 feet for features (stream, creek, road, etc.) that locate mines;
- ±80 feet for locations tied to U.S. Geological Survey topographic quad rangle sheets and ±150 feet for those tied to geologic quadrangle sheets;
- ±10 feet for mines located by survey; and
- ±250 feet (at best) for abandoned mineland portal locations.

In addition, a library was created to include digital U.S. Geological Survey 7 1/2-minute quadrangle maps, incorporating other data including gas wells, pipelines, and abandoned mine-lands projects. A set of quadrangle maps is available at nominal cost for all coal-producing counties in southwestern Virginia. A significant attribute of the Virginia program is that the digital data are accessible through industry-standard computer-aided design and geographic information systems programs. Once the mapping program is completed and all available mines have been entered in a geographic information systems database, the search for missing mines will continue with a comparison with the OSM database.

SOURCE: I. Duncan, Virginia Department of Mines, Minerals, and Energy (VDMME), personal communication 2001; and VDMME Division of Mineral Resources Maps and Publications (http://www.mme.state.va.us/DMR/DOCS/MapPub/map_pub.html).

SIDEBAR 4.2 MINE MAPS AND STORAGE, STATE OF WEST VIRGINIA

West Virginia is in the process of creating a geographic information system inventory of coal resources that incorporates both geologic information and underground mine locations, at the standard 1:24,000 scale used for U.S. Geological Survey topographic and geologic quadrangle maps. Coal resource data include outcrop locations, structural contours of coal seams, total (coal plus partings) seam thickness, and percentage of in-seam partings. Stratigraphy and coal seam correlation are accomplished using core logs, geophysical (e-logs), measured coal outcrop sections, in-mine coal sections and bottom-of-seam elevations, coal bed discontinuities, and coal analyses (heating value per pound [Btu/lb)] and percentage of sulfur, and ash).

The digital mine mapping system is in the initial stage of development. Mapping has been completed for Fayette County and the northern panhandle counties. The present focus of the project is the southern West Virginia coalfield, where the majority of current mining activity occurs. Underground, surface, and auger mines are being mapped. The West Virginia Geological and Economic Survey is concurrently mapping the location of impoundments that overlie old mine workings. Map records dating to the end of the 19th century are maintained on microfiche at two state agencies. Records for more than 100,000 mines are available for public inspection and copying.

SOURCE: N.Fedorko, West Virginia Geological Survey, personal communication, 2001.

approaches in Virginia, West Virginia, and Kentucky indicates significant differences in scope of data collection, storage, and access.

National Archive for Mine Maps

OSM maintains a National Mine Map Repository in Pittsburgh, Pennsylvania (http://mmr.osmre.gov). Some maps in the repository were originally maintained in the U.S. Bureau of Mines files and were transferred to OSM. OSM accepts maps for inclusion in the repository from various sources, including the states, and makes the archived maps available upon request. OSM has no regulation requiring the submission of maps but does have informal arrangements with MSHA to provide copies of the final map for abandoned mines. In addition, some copies of mine maps are prepared to illustrate mine ventilation systems, and these maps may not contain all the mine map information required by 30 C.F.R. §§ 75.1200 and 75.1200–1. The repository may receive such copies in lieu of the final map.

SIDEBAR 4.3 MINE MAPS AND STORAGE, COMMONWEALTH OF KENTUCKY

Kentucky maintains paper copies of maps for approximately 150,000 mines at the Kentucky Department of Mines and Minerals. Mine locations and perimeters are recorded on U.S. Geological Survey topographic quadrangles. Hard copies of the mines maps are stored at the Department of Mines and Minerals and are available for public viewing. However, Kentucky coal companies do not contribute any mine maps to the OSM National Mine Map Repository (http://mmr.osmre.gov). Recently, Kentucky's auditor of public accounts recommended that the Department of Surface Mining, Reclamation, and Enforcement, the Division of Abandoned Mine Lands, and the Division of Mines and Minerals locate, scan, and digitize underground mine maps for Kentucky (Hatchett, 2001).

Distribution of mine location information is limited by the Department of Mines and Minerals. In accordance with state law, KRS 352.480, the mine license-holder (typically the operating company), mineral owner, mineral leaseholder, or mine operator must give written consent before a mine map can be copied. Adjacent or nearby property owners may secure a copy of mine maps by filing an affidavit alleging encroachment on property outside the ownership or leasehold. The maps are copied by a private firm and may take several days. The mine maps are not stored in digital format but are generally available for mines dating from 1948 to the present.

SOURCE: Kentucky Mine Map Information Center (www.state.ky.us/agencies/cppr/dmn/mmic.htm).

The collection of maps in the OSM repository is either voluntary from the states and other sources, or results from an informal arrangement with MSHA. As a result, OSM and state records differ in many instances. *The committee concludes that the transfer of mine maps from MSHA to OSM should be better coordinated.* **The committee recommends that MSHA work with OSM and state agencies to develop a coordinated and assertive approach to collecting and archiving mine maps.**

Records of the individual states should be compared with those of OSM to ensure that both have the identical mine map database. As mine maps from active operations are submitted to MSHA and state agencies, copies should be forwarded to OSM, so the archival data can be updated. MSHA, in coordination with OSM and the state agencies, should undertake a thorough and comprehensive effort to acquire maps of abandoned mines, concurrently with the updating of maps of active mines. This process should include canvassing mining communities, land- and mineral-holding companies, and former mine employees (engineers, geologists, landmen, surveyors, superintendents, and foremen), historical societies and courthouses to acquire or copy mine maps from coal companies. The Federal Abandoned Mine Lands Fund could be a source of funding for this activity.

Storage Risks

Damage to archived mine maps can result from improper storage, natural aging of the materials used, water, fire, or other circumstances (Sidebar 4.4). MSHA requires that underground mine maps be kept in an area on the surface to minimize the danger of loss by fire or other hazards (30 C.F.R. § 75.1200). *The committee concludes that electronic data storage can reduce the risk of permanent loss or damage of mine map data and improve the ability to maintain multiple backup files both on- and off-site.* **The committee recommends that upon receipt of a paper copy of a mine map, the state or federal agency should have it scanned into electronic data files.** The original paper maps should be stored in fire- and flood-proof vaults, while electronic copies of mine maps should be stored on site with regular backup to an off-site facility.

Mine Surveying

Accurate mine maps will be critical to future site assessment for coal waste impoundments, as well as for other land-use decisions. An important component of an accurate map is the closure of the mine survey. Closure is a measure of the acceptable error within a closed-loop survey. Closure standards for surface surveys vary widely by state (Table 4.1) and type of real property (e.g., urban, suburban, or rural). MSHA regulations (30 C.F.R. § 75.1200–2) require operators of all underground mines to conduct a closed-loop survey, but do not specify the standard of closure and the distance between the last closed loop and the active face.

The committee concludes that the establishment of uniform mine surveying and mapping standards is essential to ensure that underground coal mines are accurately located with respect to other mines and surface structures, including refuse impoundments. **Therefore, the committee recommends that MSHA set standards for minimum closure error for all underground closed-loop surveys and that a closed-loop survey be maintained within a standard distance (to be determined by MSHA).**

Mine elevations are also an important component of a mine survey. The mine elevations are posted on the map adjacent to each spad. The spad is the underground equivalent of an iron pin or steel rod that marks property corners

SIDEBAR 4.4 KENTUCKY FIRE DESTROYS MINE MAPS

A fire on November 12, 1948, destroyed at least 30,000 mine maps at the Kentucky State Department of Mines and Minerals. This facility was housed in a University of Kentucky building, which also contained the Botany Department and the Kentucky Geological Survey. The Mines and Minerals Department lost, in addition to the maps, its reports, including safety records, inspection reports and recommendations. Many of these maps charted abandoned mines dating from 1884 to 1948. State geologic maps escaped the fire because they were kept at the home of the state geologist. The cost of damage was assessed at $200,000, but many of the lost records were irreplaceable.

Mine maps are now stored at Kentucky's Mine Map Information Center. This center has operated the Mine Map Repository for the Kentucky Department of Mines and Minerals for the past 28 years. The center currently houses 100,000 coal mine maps and 140,000 mine records. Although a few of the maps destroyed in the 1948 fire have been replaced, the majority of the maps in the repository date only from 1948.

SOURCE: *Courier-Journal* (Louisville), 1948; *Lexington Herald Leader,* 1948.

in a surface survey. The horizontal (x,y) location of each spad is referenced to the mine coordinate system. The z component is defined by the bottom-of-seam elevation, a critical component of mine surveying. Current surveying practice is to establish bottom-of-seam elevations by a level survey (30 C.F.R. § 75.1200–1(k)). However, older mines sometimes used the top of the seam to refer to mine elevations. Hence, caution should be used in evaluating elevations on older maps.

The primary use of elevation data is to track the flow path and potential total pressure (head) of water that may accumulate in the mine. The data can also be used for vertical location of the active mine in the framework of overlying and underlying seams, which may be actively mined or may contain abandoned mines, and for identification of undulations in the mine floor that frequently correlate with poor roof or floor conditions.

The thickness of the outcrop barrier is critical to the evaluation of blowout, blow-in, or breakthrough potential (see Chapters 3 and 6). Because the last cut is typically left unbolted, the measurement must be made remotely from the protective cover of supported mine roof. Remote measurement can be routinely accomplished with laser or sonar equipment that indicates distance by the reflection of light and sound waves, respectively. The mine engineer or land surveyor should be responsible for the accuracy in documenting the extent of the final cut. Under no circumstances should the

TABLE 4.1 Required State Surveying Closure Standards

State	Mine Surveys	Property Surveys
AL		1:10,000 for urban/commercial/high risk 1:7,500 for suburban 1:5,000 for rural
IL		ALTA* standards.
IN		None specified.
KY	None/federal	1:10,000 for urban/commercial/high risk 1:5,000 for rural
MD		1:15,000 for urban/commercial/high risk 1:10,000 for suburban 1:7,500 for rural 1:5,000 for mountains and marshland
OH**	1:5,000	1:5,000 for urban/commercial/high risk
PA		None specified.
TN	None	1:10,000 for urban/commercial high risk 1:7,500 for suburban 1:5,000 for rural
VA	Same as rural	1:20,000 1:10,000
WV		None specified.

* ALTA: American Land Title Association (1:15,000—urban, commercial/high risk; 1:10,000 suburban; 1:7,500—rural; 1:5,000—mountains and marshland)

** Data for Ohio: Mark Jones, Ohio State Board of Registration for Professional Engineers and Surveyors, personal communication, 2001.

SOURCE: C.Gillian and M.Wooldridge, Alliance Consulting, personal communication, 2001

mine foreman authorize the active section to be abandoned or pillars to be recovered until the final depth of each heading has been recorded. **The committee recommends that the mine foremen and surveyors be required to record the depth of the last cut taken to a level of accuracy to be determined by MSHA. It is imperative that any areas not completely surveyed be noted as such.**

Coordinate Systems

At least two and preferably three reference points are established on the surface, usually near the portal or shaft (30 C.F.R. § 75.1200–1 (h)) before an underground mine is opened. These surface points are tied to a coordinate system (latitude and longitude, state plane, or local) (Sidebar 4.5), so as to enable mine workings to be accurately located in the framework of surrounding surface features. The surface points are a permanent reference that can be used if the portal or shaft location has become obscured by collapse or post-mining reclamation. Subsequent mine surveying and mapping is based upon a fixed point within the mine, a point-of-beginning, that is referenced to the surface points.

The majority of active underground mines use the state plane coordinate system. The surface points are fixed surface monuments, tied to both the state plane system and at least one point referenced to latitude and longitude. Using state plane coordinates is advantageous because utilities (electric power lines, gas transmission lines), roads (federal, state, and county), and other planimetric features are typically referenced to the state plane system; and because U.S. Geological Survey topographic maps list both North American Datum 27 latitude and longitude and state plane coordinates, so mines can be located easily on commonly available maps.

The committee concludes that the variety of mapping and coordinate systems in use at present increases the potential for misinterpretation or inaccuracy in underground mine locations. **Therefore, the committee recommends that state plane coordinates or latitude and longitude and bottom-of-seam elevations as the map base reference.** However, when using latitude and longitude, the mine operator should clearly designate whether the mapping is based upon North American Datum 27 or North American Datum 83. Elevations of seam bottom, used to establish the vertical position of the mine, must be referenced to mean sea level.

Unfortunately, no uniform standard sets an appropriate coordinate system or the type and placement of surface reference points. The choice of a mine coordinate system can vary with the age of the mine, the operating company, the geographic location, or the mineral lessor. As expected, the largest variation in practice occurs in older mines (pre-1969) and small mines. Historically, local coordinate systems have been prevalent where a single entity owned the mineral rights to large, contiguous tracts. Where this practice continues, it is limited to specific properties controlled by mineral-holding companies. In these instances, a coordinate transformation between the local and state plane coordinates is given. Similarly, for small hilltop mines operating in remote areas, there was little economic justification for

SIDEBAR 4.5 COORDINATE SYSTEMS

The most commonly used coordinate system today is latitude and longitude, which are defined by reference planes based on the Prime Meridian and the Equator (Snyder, 1987). Latitude is measured in degrees north or south of the Equator, and longitude is measured in degrees east or west of the Prime Meridian.

In the United States, the state plane system was developed to provide local reference systems that were tied to a national datum. This system divides the United States into more than 100 distinct grid zones and provides an easily used, flat grid that maintains a difference between geodetic and grid distance of 1:10,000 or better. The first state plane system was developed in the 1930s and was based on the North American Datum 1927 (NAD27) (in feet). This system has been largely superseded by the North American Datum 1983 (NAD83) (in meters), although maps in NAD27 coordinates are still in use.

Mine surveyors can also use a local system based on an on-site monument. A grid is established based on direction and distance measured from the monument. These coordinate systems, because they do not take into account the curvature of the Earth, lose accuracy as the grid is extended away from the base station. In addition, these systems may be difficult to transform with other coordinate systems, especially if the on-site monument is displaced or unmarked.

SOURCE: Dana, 1999.

the time and expense associated with a closed-loop survey to tie the mine to a U.S. Geological Survey monument.

Where large mineral-holding companies control mineral rights, considerable time and effort has often focused on establishing transformation equations between local and state plane coordinates. The equation involves a fixed rotation angle and lateral offset to convert from the local system to state plane coordinates. However, in some instances a single, unique coordinate transformation is not applicable to an entire property. Thus, conversion of a particular mine map from local to state plane coordinates, requires knowledge of the geographic limits of a particular coordinate transformation. **The committee recommends that appropriate coordinate transformation equation(s) be listed on the mine map.**

Where no coordinate transformation exists, extreme care must be used in referencing maps from abandoned mines into state plane coordinates or latitude and longitude. In these situations, the transformation may be based upon aligning creeks, roads, or other planimetric features to establish the mine location. However, the surveyor or engineer must recognize that the location of these features may have been drawn freehand and not established by an accurate survey. The main objective of surveying is to depict accurately the underground mine workings; surface features are secondary. Furthermore, without survey stakes or monuments, the natural meandering of creeks over time and realignment of roads may invalidate a surface survey. *The committee concludes that surveys based on data not tied to stakes or monuments may not be accurate.* **The committee recommends that a qualifying statement accompany any coordinate transformation that is based upon the alignment of surface features.**

Surface horizontal and vertical controls should be referenced to permanent survey monuments. **The committee recommends that MSHA establish standards to improve and maintain the location of surface controls.** Monuments should be referenced to state plane coordinates and at least one monument located in North American Datum 27 or 83, or latitude and longitude (Sidebar 4.5). They should be anchored in rock and located so that they are not obliterated or obscured by reclamation and other activities associated with mine operation and closure. Monument locations should be established by closed-loop survey (with minimum closure errors consistent with underground surveying standards recommended above) from a fixed monument for which the accuracy of location is equal to that of a U.S. Geological Survey monument. The elevation of the surface monuments should be referenced to mean sea level and equal in accuracy to that of a U.S. Geological Survey monument.

Identification of Geology and Coal Seams

In some instances, companies require mine surveyors, foreman, and engineers to record geological observations or conditions in the mine floor or roof rock. However, this requirement is not universal. Roof falls, floor heave, and water or gas inflow are examples of where geologic information would commonly be noted on a mine map. As discussed in Chapters 3 and 6, the potential for subsidence is critical to quantifying the risk for slurry to enter the mine workings. Although geologic information is contained within the permit, site-specific data are important when mining adjacent to the outcrop. *The committee concludes that the geologic information contained within the mine permit documents and within company exploration and well records is useful for determining the presence and extent of potential ground weaknesses that could affect waste impoundments.* **Therefore, the committee recommends that MSHA work with the state regulatory agencies to determine which mine permit documents should be retained,**

in what form, and for how long. Such information might include the presence of fractures, faults, joints, "hill-seams" (tension cracks), and water inflow from the roof or floor.

The designation of a coal seam should be regarded as approximate, subject to verification based upon: seam elevation, seam thickness, and location of marker beds as obtained from core logs; elevations of seam bottom obtained from the mine map; or comparison with the elevations of overlying or underlying mines. Although several resources (e.g., Toothman, 1977; National Geologic Map Database's Geologic Names Lexicon at http://ngmdb.usgs.gov/Geolex/geolex_home.html) exist for designating a particular coal seam, the correct name is not always identified. For example, a coal seam may be referred to by several different names depending on geographic location, mine operator, absence of lateral geologic correlation, or economic value. This point is clearly illustrated by a few examples: In Pike County, Kentucky, the Pond Creek coal seam is also known as the Lower Elkhorn seam, which in West Virginia is synonymous with the No. 2 Gas seam. In Lynch, Kentucky, the geologic cross section shown on the U.S. Geological Survey 7 1/2-minute geological quadrangle map describes the first three above-drainage seams as the Harlan, Kellioka, and Darby. However, on the same property they were referred to as the A, B, and C by one coal company and the 180, 240, and 260 by a second company. Upon crossing Black Mountain into Virginia, these seams are known as the Wilson, B (Marker), and Taggart. *The committee concludes that coal seam names are potentially imprecise.* **Therefore, the committee recommends that coal seam names not be the sole basis for determining the vertical location of an abandoned mine.**

SUMMARY

Accurate mine maps are critical to establishing the location of underground mine workings with respect to existing or proposed coal refuse impoundments. Mine maps are the primary means by which the thickness of the outcrop barrier (horizontal separation) or overburden thickness (vertical separation) are determined. The accuracy of mines operated since the 1970s, in which surveys were made with modern equipment and closed loops, are likely to be suitable for use in the design of an impoundment. Maps for older mines may or may not be suitable. Furthermore, unrecorded final cuts may compromise the accuracy. In such cases, additional investigation of the locations of abandoned mine workings in warranted. The next chapter describes geophysical methods for such investigations.

In many instances, nonexistent, erroneous, or incomplete mine maps prevent knowing the extent, location, and depth of mined areas. **Therefore, the committee recommends that MSHA work with OSM and state agencies to establish standards for mine surveying and mapping. These should include the following:**

- **Determining surface coal outcrop locations by aerial topographic measurements, where adjacent to existing or proposed refuse impoundments,**
- **Implementing a coordinated and assertive approach to collecting and archiving mine maps,**
- **Scanning paper copies of mine maps into electronic data files upon receipt,**
- **Setting standards for minimum closure error for all underground closed-loop surveys and that a closed-loop survey be maintained within a standard distance (to be determined by MSHA),**
- **Recording the depth of the last cut taken to a level of accuracy to be determined by MSHA,**
- **Using state plane coordinates or latitude and longitude, and bottom-of-seam elevations as the map base reference,**
- **Listing of appropriate coordinate transformation equation(s) on the mine map,**
- **Adding a qualifying statement to accompany any coordinate transformation that is based upon the alignment of surface features,**
- **Improving and maintaining the location of surface controls,**
- **Determining which mine permit documents should be retained, in what form, and for how long,**
- **Avoiding the use of coal seam names as the sole basis for determining the vertical location of an abandoned mine.**

5

Technologies for Locating Mine Workings

Locating underground mine workings in the absence of mine maps is not impossible, but it can be expensive and time consuming. Drilling enough vertical holes on a 3-foot grid will locate virtually any mine workings, but, in addition to the expense, it damages the Earth's surface and creates conduits for underground fluids, including pollutants that can degrade groundwater. Because of the time and expense associated with extensive drilling, remote sensing and geophysical methods have been employed to search for abandoned coal mines (e.g., Branham and Steeples, 1988; Miller and Steeples, 1991, 1995). The objective of geophysical surveys is to provide descriptive information about the physical characteristics of a three-dimensional volume of earth material, including the presence of voids. Because no geophysical technique is capable of performing optimally under all geological and topographic conditions, multiple geophysical techniques may be necessary to reduce the probability for error to an acceptable level. While these methods have proved successful in some cases, drilling is still necessary to confirm interpretations of geophysical and remote sensing data (Table 5.1). In addition, the absence of evidence of a mine is not evidence of absence of a mine, and there are many opportunities for error in the modeling and geophysical surveys needed to detect voids.

To give regional context to a local area, geophysical surveys begin with baseline information. Consequently, the area surveyed with geophysical methods is commonly several times larger than the planned impoundment, particularly for the less expensive methods such as magnetic surveys.

Furthermore, during the planning and interpretation stages, geophysical surveys and data collection should be accompanied by geophysical modeling using the tools of physics, geology, engineering, and mathematics (NRC, 2000a). Today, modeling software that runs on personal computers is available for all of the geophysical techniques. Moreover, in-field modeling with a personal computer can often be of use in making evaluations during

TABLE 5.1 Geophysical Methods and How They Work

Method	Principle	Typical Measurement	Physical Property Measured	Interpreted Parameters
Airborne sensing	Detects reflected electromagnetic radiation	Aerial photography and remote sensing in several spectral bands	Spectral dependent reflectance of electromagnetic radiation	Geologic lineations, variations in vegetation, surface disturbances
Electrical and electromagnetic	Detects current flow in subsurface materials	Currents, voltages, spatial locations	Electrical resistivity	Depth, earth material resistivity, porosity, inferred fluid chemistry
Ground-penetrating radar	Transmits radio waves in the 10- to 500-MHz band into subsurface and detects returning reflected waves	Distance, wave arrival times, and wave amplitude	Dielectric permittivity, electrical resistivity, magnetic susceptibility	Shallow interface depth and geometry, electromagnetic wave speed, electromagnetic wave attenuation
Magnetics	Detects local variations in the Earth's magnetic field caused by magnetic properties of subsurface materials	Proton precession frequency	Magnetic susceptibility	Geometry and magnetic susceptibility of local subsurface features
Microgravity	Detects localized minute variations in the gravitational field of the Earth	Displacement of a gravitational-force-sensitive mass	Mass density	Depth, geometry, and density of local subsurface features
Seismic methods	Source of seismic waves provides sampling of elastic properties in a localized volume of the Earth	Distance, wave arrival time, and wave amplitude, different wave types	Speeds of compressional, shear, and surface waves attenuation of these waves	Interface depth and ; geometry, elastic moduli, location of faults

SOURCE: Modified from NRC, 2000a.

the data collection process, including changing the extent and emphasis of a survey.

The appropriate strategy for finding subsurface cavities depends on the size of the target. That size determines the spacing of geophysical measurements, which should be sufficient to define subsurface conditions. Decreasing density costs less but may increase the probability of error in the detection of voids. To achieve a suitable detection probability, any site investigation should work from a regional to a local to a site-specific approach; techniques should progress from large-scale and noninvasive (e.g., aerial photography, remote sensing) to detailed site data (e.g., drilling, borehole geophysical methods). Holes should not be drilled until the investigators know what they are looking for, and measurements should not begin without an adequate understanding of the surrounding geology (R.Benson, Technos, personal communication, 2001).

The inclusion of microprocessors in modem geophysical instrumentation has made on-screen, menu-driven geophysical data collection easy for those who have limited geophysical training. Problems can arise when the data are collected with inadequate spatial sampling density or with inappropriate techniques, and these problems can be compounded when individuals with inadequate geophysical education and experience attempt to interpret the data. Even if the data have been collected correctly, inappropriate data processing and interpretation can introduce many pitfalls. Furthermore, no single noninvasive technology works everywhere all the time. Consequently, an interdisciplinary approach is needed to interpret and to integrate multiple geophysical data sets with local geological and engineering data.

This chapter offers an overview of the site characterization techniques that can be utilized to locate abandoned underground mines. A more detailed description of these techniques is available in a guide for selecting surface geophysical methods issued by the American Society for Testing and Materials (ASTM, 1999) and in Appendix E.

DRILLING

There are several ways to drill shallow holes into the Earth (e.g., NRC, 1994). The simplest is done using an auger that brings soil material to the surface much like a drill bit boring in wood brings wood cuttings to the surface (Hynes, 1995). Augers cannot penetrate solid rock; they are used primarily to drill in soil, unconsolidated sands and clays, and soft shales. Only the shallowest of coal beds can be reached with auger drills. Most modern drilling at depths of more than 100 feet or so is done using rotary methods in which air, water, or drilling mud (often containing high-density barite) is circulated down the center of the drill pipe and back up the outside of the drill pipe to help move the cuttings to the top of the hole. If solid pieces of rock rather than cuttings are required for analysis, the investigator must resort to core drilling, which is usually more expensive than simple rotary drilling.

REMOTE SENSING

One of the principal uses of remote sensing in coal-related activities is to detect linear features, such as faults and fracture zones, on scales from miles to tens of miles. Delineation and evaluation of such regional linear features is necessary as part of the permit application process for a new impoundment. Remote sensing provides observations of the Earth's surface and shallow (less than 3 feet) subsurface to complement information from mapping and geophysical methods. Because the measurements are commonly made from a moving airborne platform, a large area can be examined quickly and cost-effectively.

For remote sensing to be useful in evaluating subsidence-related problems, spatial resolution on the order of 3 feet or less is required. Measurements from low-flying aircraft, or even from unmanned aircraft, can provide such high-resolution measurements and photography. For additional information on remote sensing applications, see NRC (1995) and Watson and Knepper (1994).

The oldest form of remote sensing is aerial photography. Historic black-and-white aerial photos, often dating from 1930s Works Projects Administration files, are sometimes available for comparison with modern photography. Modern scanning and digital registration techniques can be used to enter these old photos into a geographic information system format. Historic photos may indicate shafts and other facilities that have been abandoned and covered but not properly plugged, and are also useful for analysis of drainage patterns, detection of geologic conditions, such as landslides and debris flows, and signs of vegetative change, including stress from pollution or from shallow voids.

Multispectral scanning samples the field at several different parts of the electromagnetic spectrum (including outside the wavelength range visible to humans) and digitizes the data for later analysis. The spectral bands sampled are selected to provide the most sensitivity to features that can characterize problems of interest. Individual portions of the spectrum are analyzed with digital signal processing techniques and the results are compared with statistical or physical models or to information extracted from specimen-based laboratory measurements.

For regional analysis (tens to hundreds of square miles), the 100-foot resolution satellite-based thematic mapper is a useful tool. It employs six different bands in the reflected spectrum, along with a 400-foot resolution thermal channel. Because of the spatial resolution limitations, the thematic mapper is not useful for the detailed analysis needed for localized problems. Regional studies can also benefit from archival radar and photographic data from national surveys, available from the U.S. Geological Survey's Earth Resources Observation System Data Center. The IKONOS high-resolution satellite, which offers 3-foot resolution, also is useful for mapping and surveying.

For localized studies, aircraft systems that use additional spectral channels and better spatial resolution can be employed. Reflectance data have been used to distinguish among geologic units, to detect geologic structures, and to trace linear fractures. They have also been used to indirectly infer lithologic and soil information in vegetated areas based on empirical relations between vegetation and geological conditions. Thermal infrared data can be used to detect exothermic reactions, such as underground coal fires, and to find hydrological features such as springs (NRC, 2000a).

GEOPHYSICAL METHODS

Noninvasive active geophysical methods used to search for coal mine voids employ artificial electrical, electromagnetic, or mechanical energy to examine the shallow subsurface of the Earth (Sharma, 1997; Ward, 1990). In contrast, passive geophysical techniques measure some natural physical parameters of the Earth, such as minute variations in the Earth's gravitational field (see Dobecki and Romig's 1985 review).

Physical parameters measured directly during shallow geophysical surveys include: electrical, thermal and stress fields, gravitational and magnetic fields, electrical conductivity, elastic (i.e. seismic) properties, transparency to and polarizability of electromagnetic waves, and natural gamma radiation. These measurements can then be used to infer the permeability, porosity, chemical constitution, stratigraphy, geologic structure, and various other properties of a volume of material near the Earth's surface, including the presence of voids. Geophysical methods can be used to guide exploratory drilling programs, but they cannot be expected to eliminate confirmation drilling.

The geophysical methods used and how they are applied vary according to a project's objectives, resolution requirements, budget, and geological situation. For example, seismic methods are sensitive to the mechanical properties of earth materials but not to the chemical makeup of these materials and the fluids they contain. In contrast, electrical methods are sensitive both to fluids and to magnetic or electrically conductive materials. Usually, multiple geophysical methods offer better answers than any individual method. The difficulty (and sometimes the success) of geophysical surveys is affected by topography. For example, it is easier to collect, process, and interpret geophysical data in the agricultural fields of Illinois than in the steep valleys of West Virginia.

In addition to topography, vegetation and cultural features—such as buildings, roads with traffic, and fences—can be barriers to geophysical surveys. Furthermore, because of the need to obtain regional geophysical background information, the area needed for a geophysical survey may greatly exceed that owned or leased by a mining company, in which case rights of ingress and egress for geophysical measurements may be a serious issue.

Electrical Resistivity and Electromagnetic Methods

Resistivity techniques sense the electrical properties of the material through which a current passes. Electrically conductive contaminants can be tracked using resistivity methods. For example, resistivity would be expected to be more effective for finding mine workings full of polluted water than for detecting mine workings full of air.

Under some conditions, these methods can be used to find geological faults and buried valleys—but usually not with the precision of seismic reflection techniques. Resistivity surveys are usually cheaper than seismic surveys. Multichannel electrical cables similar to seismic cables have recently been developed to increase the flexibility and the rate of resistivity data collection. Electrical and electromagnetic survey data interpretation often involves mathematical inversion, producing a model that fits the data (e.g., Ellis and Oldenburg, 1994; Zohdy, 1989). Electromagnetic methods have partially replaced resistivity surveys because equivalent information is obtained faster and without inserting electrodes into the ground.

Electromagnetic methods include active methods in which an electromagnetic signal is induced in the ground by human activity, and passive methods in which natural variations in the electromagnetic field of the Earth are analyzed (Appendix E). The induced polarization method is related to the resistivity method, except that the Earth's delayed response to an induced current is analyzed. The spontaneous potential method measures the natural voltage in the Earth resulting from electrochemical activity. If water in mine workings is reacting with its surroundings, this might be detectable with a spontaneous potential survey. Water in motion through fractured and porous media produces a "streaming potential," which is useful in detecting leaks in dams. This technique might also be useful in looking for mine works in which water is flowing.

Active electromagnetic methods have become more popular in near-surface geophysical applications (Appendix E). The theoretical basis for, as well as practical background for, electromagnetic methods is provided in McNeill (1990). These methods have a major advantage over direct-current resistivity because they do not require placing electrodes in the ground. Indeed, the surveys can sometimes be conducted from low-flying aircraft. A recent development in airborne electromagnetics offers the advantages of increased surveying speed and access to polluted, dangerous, or inaccessible areas via small (maximum dimension 3 to 6 feet) unmanned aircraft. However, airborne surveys also have disadvantages, including limited separation between the source and receiver coils and a higher noise level caused by the movement of the coils through the Earth's magnetic field (Blakely, 1996; Nabighian, 1988, 1991).

Potential Field Methods

Buried metal objects such as steel drums are often found with magnetometer surveys in which measurements with precision of one part in 50,000 of the Earth's total magnetic field are made. Although data precision and collection rates continue to improve, magnetic surveying is a relatively mature science. In the future, vector recording of the magnetic field instead of the commonly used total field could be useful. Because coal is relatively nonmagnetic, the removal of coal does not alter the magnetic field very much. Consequently, magnetic surveys are not commonly useful in finding underground mine workings. They could be useful, however, in detecting old cased wells or mine workings that contain metal pipes, cables, rails, or equipment. Figure 5.1 shows the size of a magnetic anomaly that is typical at various distances for common metallic items such as tools and vehicles.

Magnetic gradiometry consists of taking simultaneous readings from two magnetometers spaced a few inches to several feet apart and analyzing the difference (the magnetic gradient). Magnetic surveys are also useful in mapping faults, locating magnetic bodies, and estimating the depth to

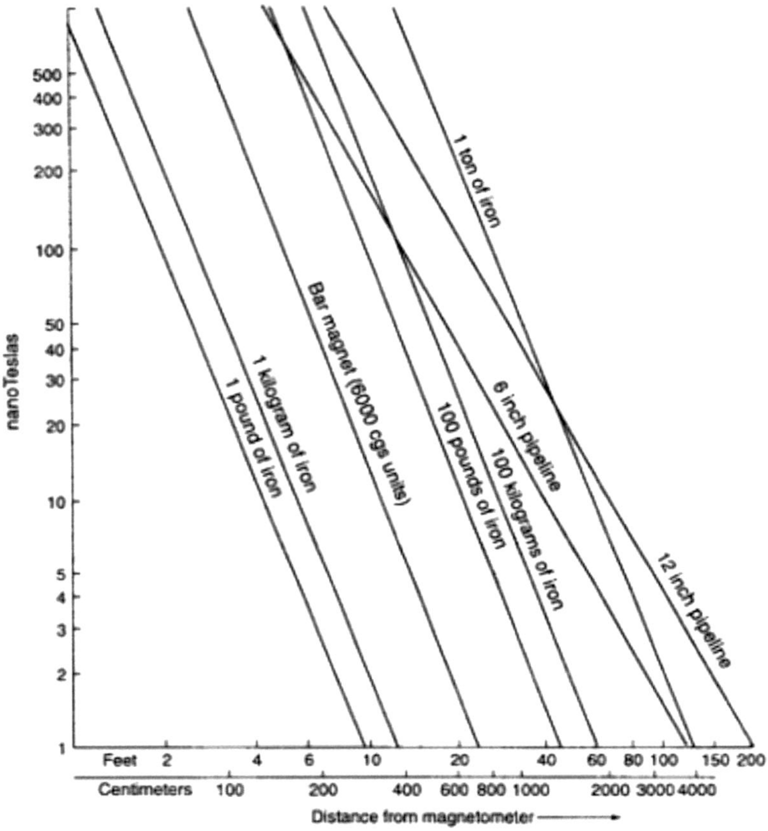

FIGURE 5.1 Example of magnetic anomaly at various distances for common metallic items such as tools and vehicles. Reprinted with permission from Breiner, 1973. Copyright 1973 by Geometries.

magnetic earth materials. Such surveys are used to detect variation in the magnetite content of rocks and unconsolidated materials, so they can detect changes in some types of igneous rocks and other geologic structures. They are used also at contaminated sites to measure the perturbation of the Earth's magnetic field caused by buried ferrous metal objects such as steel drums, the ferrous metal waste in landfills, and iron pipes (e.g., Roberts et al., 1990a).

Microgravity surveys measure minuscule changes in the gravitational field of the Earth using gravity-meters with a sensitivity of 1 microGal (where 1 Gal is a gravitational acceleration of 1 cm/sec/sec). Readings are made along a profile line or on a grid with typical spacing of 1 to 100 meters (3 to 330 feet). The sensitivity of microgravity measurements is one part per billion in comparison to the Earth's gravitational field. A map of gravity anomalies reflects the lateral density contrasts detectable after removing all known effects that can cause changes in gravity, such as, tide, instrument drift, elevation and latitudinal variation, and terrain.

For near-surface geophysical exploration, microgravity surveys sometimes are used where a high contrast in density occurs between bedrock and overlying alluvium, or between air and rock in a mine. Minute gravity anomalies can be caused by artificial features such as trenches, tunnels, disposal containers, and incipient subsidence problems (e.g., Roberts et al., 1990b; Yule et al., 1998), as well as by geologic features such as cavities, faults, folds, dipping layers, and lateral intralayer heterogeneity.

While microgravity methods could be applicable in finding shallow air-filled mine workings, they would not be the first choice for finding water-filled mine workings because the density contrast between the missing coal and the water that replaced it is too small. Figure 5.2 shows the calculated gravity anomaly at the Earth's surface above an air-filled 20-foot-diameter horizontal, cylindrical mine entry.

Near-Surface Seismic Methods

Seismic research has met with limited success when conducted to detect cavities resulting from abandoned subsurface coal mines (Fisher, 1971; Hasbrouck and Padget, 1982), salt-solution mining (Cook, 1965), lava-flow tunnels (Watkins et al., 1967), and natural caverns (Rechtien and Stewart, 1975). Most researchers using seismic techniques for cavity detection cite three phenomena as evidence of a cavity: free oscillations or resonance of the cavity walls, anomalous amplitude attenuations, and delay of arrival time (Cook, 1965; Fisher, 1971; Godson and Watkins, 1968; Robinson and Coruh,

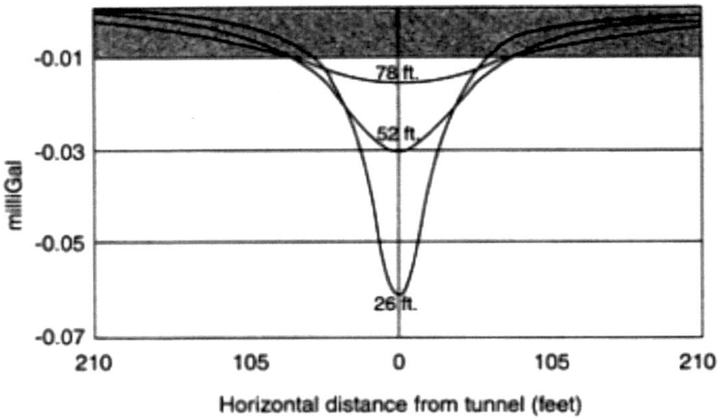

FIGURE 5.2 Calculated theoretical gravity anomaly for three air-filled 20-foot diameter horizontal cylindrical openings. The three curves represent calculations for depths of 26, 52, and 78 feet to the center of the opening, with the largest negative anomaly associated with the shallowest opening. The background and instrumental noise for microgravity methods limit the absolute value of useful anomalies to about 0.01 mGal. If the cavity were water-filled instead of air-filled, the anomalies would only be one third as large.

1988; Watkins et al., 1967). In addition, some success in locating water-filled coal-mine cavities at depths of less than 50 feet has been reported using high-resolution *P*-wave (congressional, i.e., sound waves) reflection seismology techniques (Branham and Steeples, 1988; Miller and Steeples, 1991) in which absence of a normally strong coal-bed seismic reflection indicates a mined-out coal bed.

Cook (1965) found that seismic energy transmitted through a cavity and reflected from a deeper horizon gives rise to a seismic amplitude "shadow." Most of the seismic research on coal-mine detection has involved *P*-wave refraction seismology or *S*-wave (distortional, i.e., shear waves) reflection seismology. Significant improvements have been made since 1980 in near-surface *P*-wave seismic-reflection techniques (Hunter et al., 1984; Steeples and Miller, 1990), shallow-seismic refraction interpretation (Lankston and Lankston, 1986; Palmer, 1980), and surface-wave methods (Park et al., 1999; Stokoe et al., 1994; Xia et al., 1999). Surface-wave phase anomalies might be employed to detect near-surface voids (Rechtien and Stewart, 1975). Shallow *S*-wave reflection survey results are reported in the literature (e.g., Goforth and Hayward, 1992; Hasbrouck, 1991), but separating the *S*-wave reflections from the surface waves that often appear on seismograms at the same time is a problem. Seismic shear waves may be useful for cavity detection because they will not propagate through voids or water-filled cavities. Shear wave reflections have also been used to evaluate the resources of a shallow coal seam (Hasbrouck and Padget, 1982).

Geophysical tomography is conceptually and mathematically identical to medical tomography in which three-dimensional X-ray imaging from within the human body is accomplished by computed axial tomography (CAT scan). The technique uses measured travel times or signal strength of many geophysical ray paths through a volume of earth material. Seismic tomography has been used to examine Earth's interior from scales of a tens of feet to thousands of miles (e.g., Clayton and Stolt,1981; Humphreys et al., 1984).

Future seismic applications that merge *P*- and *S*-wave refraction information may be useful (Hasbrouck, 1987). By combining *P*-wave and *S*-wave velocities with density readings obtained from gravity surveys or borehole density logs, one can measure the elastic parameters of rocks. When densities and velocities are known, Poisson's ratio, Young's modulus, and the shear modulus can be calculated. When these elastic constants are known, rock types can often be identified and an estimate of pore-space fluid content usually is possible (Domenico and Danbom, 1987).

New opportunities for three-component recording and multimode analysis are a result of decreasing cost and increasing capabilities of seismic hardware designed to collect and process high-resolution, near-surface seismic data. The seismic wave types, generally discarded by classical seismic reflection surveyors during the processing, analysis, and interpretation of data, contain information about the upper tens of feet of the Earth. The capabilities of seismic methods involving depths shallower than 100 feet can be extended by analyzing the near-surface broadband seismic wavefield using three vector components rather than one and by examining multiple types (modes) of seismic waves instead of just *P*-waves.

The principles of in-seam seismic transmission and reflection surveys can be applied to estimate the presence and location of faults (Buchanan et al., 1981; Greenhalgh et al., 1986) and air- or water-filled or collapsed mine workings (Mason, 1981). In-seam seismic surveys are typically performed in panels surrounding blocks of coal prior to long-wall mining operations. Seismic-wave transmission surveys are set up to test the transmissivity of the coal seam by deploying seismic sources along one face of a coal panel and placing geophones along the opposing face. If disturbances are inferred from the transmission experiment, a seismic reflection survey may be used to estimate their locations. Brentrup (1970) summarizes more than 200 in-seam seismic surveys, reporting a success rate of 66 percent for reflection and 83 percent for transmission surveys. Since the 1960s, in-seam seismic methods have been applied in coal mines to find disturbances in the coal seams that may pose problems for mining (Schwaetzer, 1965). Disturbances include faults that offset coal seams and reduce production rates of longwall mining operations, abandoned mine workings that pose a general risk to underground activities, and methane zones in which mining activity may trigger explosions. A more extensive discussion of in-seam seismic methods is included in Appendix E.

Faults with vertical displacement of one or more seam thicknesses are good reflectors of in-seam seismic waves because of the contrast between the coal and the bedrock material. Water- or air-filled cavities are even better reflectors. In the case of collapsed mine workings, in-seam seismic waves may be scattered rather than reflected, so the reflected waves may not be as coherent as those reflected from a fault or a cavity. Nevertheless, the amplitude of the high-frequency transmitted in-seam seismic waves could still be sufficiently decreased to infer the presence of collapsed workings. In-seam seismic techniques can be applied to existing and planned coal-waste impoundments as long as the appropriate coal seam is accessible underground. This includes cross-hole tomographic methods, which have proven successful in other applications. The overburden material and surface topography will have no effect on the success of the experiment. Many improvements in seismic equipment and analysis techniques introduced in the last two decades enable a success rate greater than that noted by Brentrup (1970).

The advantages of in-seam seismic technology lie in the two-dimensional propagation of seismic waves in coal seams, and often in accessibility to a coal seam on both sides of an assumed disturbance. In-seam seismic methods allow the use of higher frequencies and broader seismic bandwidths than surface seismic methods, and these offer better resolution of features of interest. The high-frequency waves are concentrated within the coal seam whereas lower-frequency waves are present within as well as outside the seam (Gritto and Dresen, 1992). The advantage of absence of surface seismic noise such as wind and road traffic may be countered by mining activity, which might have to be interrupted during the seismic survey within a distance of 0.5 to 1 mile from the survey. However, the problems associated with surface topography are mostly negated.

Ground-Penetrating Radar

In many places, ground-penetrating radar has become the method of choice for exploring the upper several feet of the Earth's subsurface (e.g., Daniels, 1996; Davis and Annan, 1989). Ground-penetrating radar employs a source of microwave radiation that is radiated into the Earth via an antenna at a known time. From the time measured for the wave to echo back to a receiving antenna at the surface, one can calculate the depth to various layers in the Earth, once an accurate wave-propagation velocity has been measured. Velocity determination is a critical factor and can be done by measurements on samples, by one-way travel time in a borehole, or by fitting a least-squares hyperbola to an observed-travel time versus horizontal-distance curve (Tillard and Dubois, 1995). In geometrical concept, ground-penetrating radar and seismic reflection are similar. Ground-penetrating radar data can be displayed in a format identical to that used for seismic sections. However, the Earth environments in which the two techniques perform optimally tend to be mutually exclusive. Ground-penetrating radar works best in dry media in the absence of clays or other electrically conducting earth materials because electromagnetic radiation cannot penetrate into electrical conductors (e.g., moist clay). In contrast, seismic waves transmit well through moist clays but are rapidly attenuated in dry sand. Ground-penetrating radar imaging works at depths of 30 to 60 feet under favorable conditions, but may fail at depths of less than a 3 feet if clays or other conductive materials are present near the Earth's surface.

Nuclear Magnetic Resonance

Magnetic resonance imaging was developed for medical diagnosis as an outgrowth of physicists' nuclear magnetic resonance experiments (Knight et al., 1999). Proton nuclear magnetic resonance, which responds to the state of hydrogen nuclei in the ground, is of interest to geoscientists. Proton nuclear magnetic resonance might be used to detect water-filled cavities because of its sensitivity to hydrogen nuclei. However, owing to the immature state of research in this field, referring to the geophysical use of nuclear magnetic resonance as "imaging" may be premature (Appendix E).

Radiometric Techniques

Radiometric methods measure emissions from radioactive isotopes. The techniques can be used to sense the presence of radioactive contaminants or to explore for radioactive ores. Specific isotopes that occur within 3 to 6 feet of the Earth's surface can be identified by spectral gamma methods, which may also be useful for locating radioactive hazards, such as natural radon gas sources. Nielson et al. (1990) review natural gamma and radon emanation methods.

Methods that measure anomalously high radon concentrations show promise for identifying abandoned underground mine workings. Research using this method was pioneered by geologists looking for buried geologic features such as faults (Fleischer and Mogro-Campero, 1979; Heirendt, 1988), and the method has since been adapted to locate abandoned coal mines (Misquitta, 1989). Radon is thought to be concentrated in the voids left by mining, then released to the surface by way of subsidence fractures that result from mine collapse. The method is based on the concept that the alpha particle emission that occurs during the radioactive decay of radon will leave impressions on high-density plastic detectors and that the number of impresssions on the detectors can be directly correlated with standardized radon concentrations (Figure 5.3). Anomalously high radon concentrations may correlate with the portion of the profile above the mine voids. Further refinement and development of this method could lead to a cost-effective, noninvasive screening method for detecting abandoned underground mines.

Measurement of soil gases or gases that emanate from the ground is commonly used to detect buried wastes or containers. For example, photoionization detectors are hand-held instruments that detect gasoline derivatives or additives from underground, leaking storage tanks. Natural gas utilities and transmission companies also use similar instruments ("sniffers") to detect gas leaks. The technical literature is sparse regarding the innovative use of gas monitoring to detect old mine workings in coal seams.

In the case of underground coal mines, the cracks induced by mine collapse and subsequent subsidence create fractures that can allow for accumulated gases to escape from mine workings to the overlying soil or ground surface. Methane gas is often associated with coal seams, and thus, portable gas detectors could be useful as a screening tool to identify where mine voids with accumulated gas are connected to the surface by natural or subsidence-induced fractures. Detection of anomalous gas concentrations may not alone be indicative of mine voids, but this method could prove useful as a screening tool for a drilling or geophysical investigation of a specific area. The method could reduce costs while helping to focus more

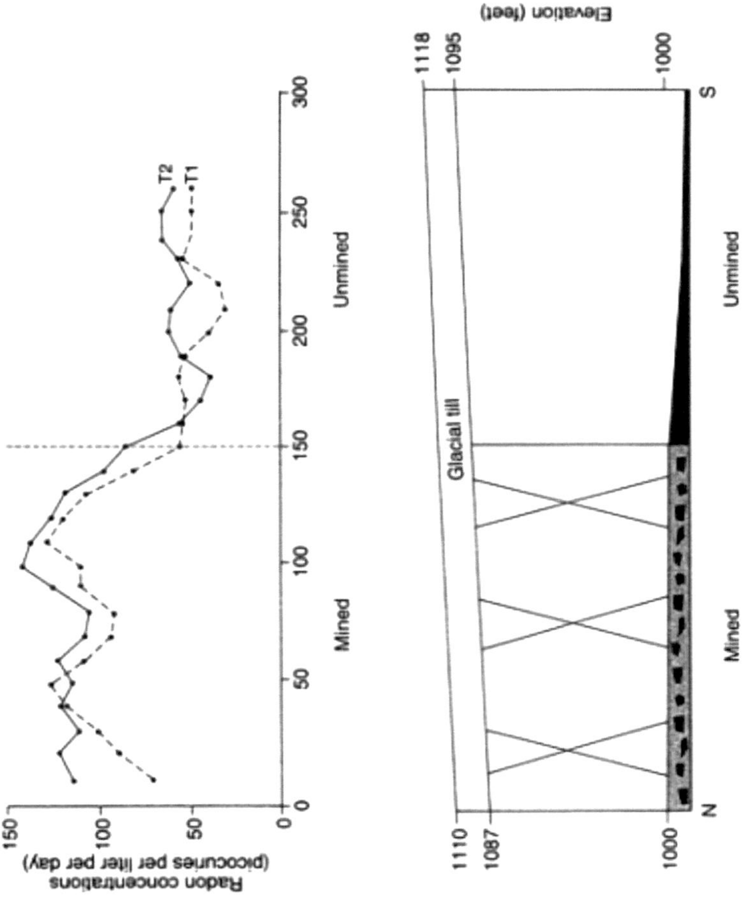

FIGURE 5.3 Radon concentration along profile including abandoned mine. From Misquitta, 1989.

quantitative methods such as drilling, and may aid in the success in detecting unmapped, mined-out areas adjacent to waste impoundments.

Borehole Geophysical Methods

Borehole geophysics is done by lowering a long sensing tool (called a sonde) attached to a cable into the drill hole (Daniels and Keys, 1990; Hearst et al., 2000) (Appendix E). The geophysical information is relayed to the surface via the cable, and the data are recorded at the surface for later analyses. The resulting data provide a plot—a geophysical log—of various geophysical parameters as a function of depth in the hole.

To detect underground mine workings, borehole geophysics can complement drilling by examining the area around the borehole to a radius of 3 feet or more. Also, if two or more boreholes are available, cross-hole tomographic analyses can be performed with several of the borehole geophysical methods. Borehole geophysical measurements offer the best resolution and decrease the effects of near-surface signal attenuation, formation heterogeneity, and some types of "noise." Many near-surface geophysical technologies obtain a degree of ground-truth from such borehole measurements. Small-diameter versions of the logging tools used in the petroleum industry have been developed for near-surface investigations.

Near-Surface Geophysical Research

Near-surface geophysical methods should have a bright future in the coal mining industry. Many near-surface geophysical techniques are still developing rapidly; though limitations imposed by steep terrain must be addressed. Today, for example, ground-penetrating radar data are collected using a single receiving antenna, but using multiple antennae could enhance ground-penetrating radar capabilities in much the same way that the seismic reflection method was revolutionized by common-midpoint surveying in the 1960s.

Theoretically, the potential is great for widespread use of seismic techniques for detection of voids, such as underground coal mines. However, currently available two-dimensional seismic reflection and refraction methods have met with limited success. Extending the usefulness of the seismic method for void detection will require new, state-of-the-art techniques that utilize more of the wavefield than seismic *P*-wave reflection alone.

A large portion of the information contained within shallow, high-resolution seismograms is not used or emphasized in standard two- or three-dimensional reflection surveys. Cavity detection is a fundamentally different problem than those addressed in standard petroleum exploration. Although the reflected portion of the wavefield may yield the highest resolution information about void location, other portions of the wavefield may provide key constraints in the detection and interpretation phases of a survey. Thus, a variety of different processing, modeling, display, and interpretation methods should be investigated to determine whether it is possible to exploit seismic wavefields uniquely altered by the presence of a shallow subsurface tunnel.

No seismic method appears to be uniformly applicable in the highly variable near-surface geology of different tunneling environments. Each method will probably be of use in some specific geological environment (Tables 5.2 and 5.3). Sidebars 5.1, 5.2, and 5.3 illustrate case histories where seismic reflection has been used to locate mine workings. The mine workings cause a shadow effect, which decreases signal strength for reflection from layers below the coal (Sidebar 5.1). The void presence in Sidebar 5.2 causes the coal bed reflection to disappear. At present the immense computational resources required limit the full waveform inversion of both seismic and ground-penetrating radar data to small data sets. When computing costs have decreased sufficiently, these inversions may become commonplace. One caveat, however, is that the inversion process treats noise with the same reverence that it treats data. When noise is present in shallow seismic or ground-penetrating radar data, a data-inversion routine may produce artifacts related to its attempt to invert the noise.

Automation could improve very near-surface geophysical methods— from model airplanes carrying microsensing devices to robots roving the ground over hazardous or polluted areas. Automatic emplacement of geophones (Steeples et al., 1999) may significantly improve cost-effectiveness of near-surface seismic surveys. All geophysical techniques could benefit from improved precision, resolution, and bandwidth. Data processing would benefit from faster and more robust methods, especially if the ambiguities and uncertainties that plague data interpretation could be decreased. By combining robotics, automation, expert systems, and miniature aircraft it may be possible to decrease costs and improve productivity. The efficient and timely transfer of technology from developers to users and potential beneficiaries could be enhanced through continuing education programs.

TABLE 5.2 General Applicability of Geophysical Methods to Coal Waste Impoundment Problems

Method	Geologic Mapping	Hydrogeology Characteristics	Fracture Detection	Cavity Detection	Coal Mine Detection	Soil Mechanics
Airborne sensing	2	3	2	3	3	3
Electrical resistivity	1	1	2	1	2	3
Electromagnetics	1	1	2	1	2	3
Ground penetrating radar	1	1	2	1	2	2
Induced polarization	2	3	2	3	3	3
Magnetics	2	3	2	3	3	3
Microgravity	2	2	2	1	1	2
Seismic reflection	1	2	1	2	1	1
Seismic refraction	1	1	2	2	2	1
Seismic surface waves	2	2	1	1	2	1

NOTE: 1= primary applicability; 2= secondary applicability; 3= limited or no general applicability.

This table indicates the typical relative applicability of these methods but should not be used as a basis for definitive planning or contracts. There are many exceptions to these generalities.

SOURCE: Modified from NRC, 2000.

TABLE 5.3 Geophysical Techniques for Void Detection

Technique	Material in Cavity	Depth of Use	Cost	Advantage	Disadvantage
Airborne sensing	Air	Vegetation roots	Low	Fast	Low resolution
Electrical resistivity	Water	656 feet or less	Medium	Good if water-filled	Poor in air-filled
Electromagnetics (active)	Water	656 feet or less	Low	Fast, easy data collection	Poor in air-filled
Ground-penetrating radar	Air	33 feet or less	Low	Fast, easy data collection	Clay wipes out signal
Induced polarization	Water	656 feet or less	Medium	Unknown	Success undemonstrated
Magnetics	Scrap metal	164 feet or less	Low	Easy data collection	Need metal in mine
Microgravity	Air	66 feet or less	Medium	Simple theory	Geologic noise
Seismic reflection	Air or water	656 feet or less	High	Produces image	Intensive data processing
Seismic refraction	Air or water	656 feet or less	Medium	Easy data processing	Indirect detection
Seismic surface waves	Air or water	98 feet or less	Medium	Easy interpretation	Depth of penetration

SIDEBAR 5.1 CASE HISTORY: COAL VOID SCHEMATIC

Seismic reflection methods have been used successfully in a few cases to detect old coal mine workings. This diagram depicts schematic seismic-reflection data superimposed on a hypothetical geological cross-section containing a coal mine void. A coal seam usually produces a strong seismic reflection from its top and bottom interfaces with surrounding rock, such as shale. The reflections from the top and bottom of the coal are represented by waves with blackened peaks that can be followed by eye in a coherent fashion from one seismic trace to another. While the coal represents a strong seismic reflector, the absence of coal (i.e. the mine void) results in the absence of the strong reflection as illustrated by the two seismic traces that pass through the void without producing a reflection.

The seismic data are processed and displayed such that each seismic trace represents seismic-wave motion as a function of time, as if the seismic source (such as a small explosion) and the seismic receiver (a geophone) were located at the same point on the Earth's surface. Hence, the reflections with blackened peaks occur at a time on the seismogram that represents travel from the Earth's surface downward to the coal seam and then back to the Earth's surface.

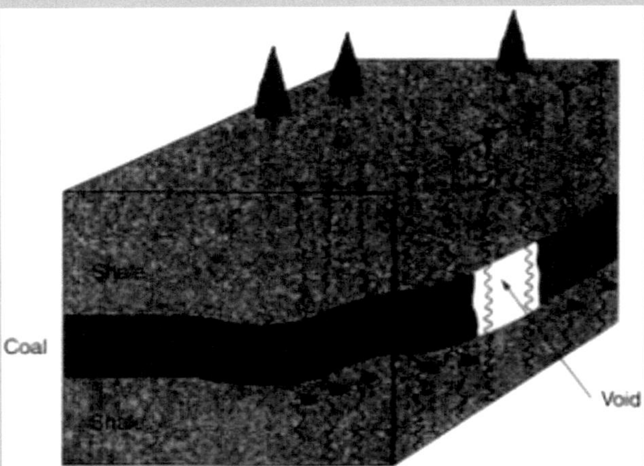

Schematic seismic-reflection response to a coal-mine void.

SIDEBAR 5.2 HIGH-FREQUENCY SEISMIC CASE HISTORY

In some cases it is possible to infer the presence of mines from decreased signal strength of seismic reflections from layers beneath the coal, which in this instance is about 550 feet deep. The vertical axis is in seconds of two-way reflection time, and the coal reflection is present at about 0.13 seconds. In contrast to the Pittsburg, Kansas, example (Sidebar 5.3), it is not possible to distinguish individual rooms and pillars in this figure. It is possible to see the general location of the mined area and to define the mine boundaries to within about 65 feet (Waters, 1987). The location of the mine near the center of the seismic section is indicated by a "faded" area except where the coal reflection is present. The coal reflection also has a lower frequency appearance in the mined area than in the unmined area.

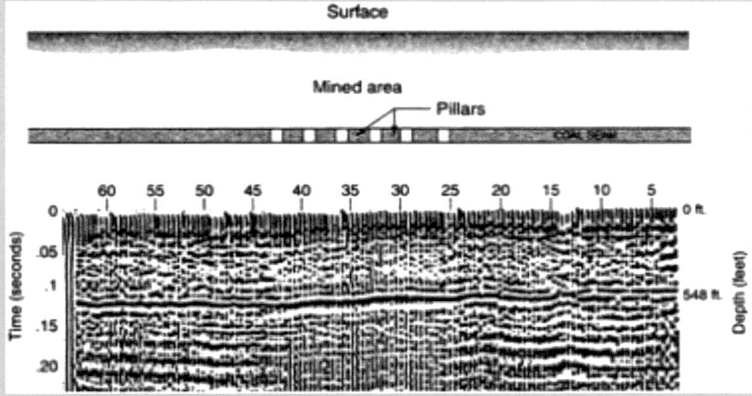

Seismic detection of mine works at 548-foot depth. Courtesy of CONOCO, Inc.

HYDRAULIC TESTING

A potential technique that may aid in determining the extent of a coal outcrop barrier or coal seam is to test a questionable area hydraulically. This is accomplished by conducting hydraulic packer tests in boreholes drilled into the coal (Harlow and LeCain, 1991; Minus, 1993). Tested intervals where the permeability of the coal is significantly higher than the statistical range for confined coal seams would be suspected of having void space within the smaller surface volume of unconfined coal. The increased stress could result

SIDEBAR 5.3 CASE STUDY: CAVITY DETECTION

The principle illustrated in the schematic diagram was used successfully to detect abandoned mine workings in an industrial park at Pittsburg, Kansas (Branham and Steeples, 1988). In this case, the coal seam was about 3 feet thick at a depth of 33 feet. The first two coherent blackened peaks on the seismic section represent seismic refractions rather than reflections. The third and fourth blackened peaks represent the coal reflection, which is absent where the seismic survey passed directly above the abandoned mine workings. The geological cross section above the seismic section shows the geological interpretation that was supported by two boreholes, one of which hit a mine void and the other, the coal seam.

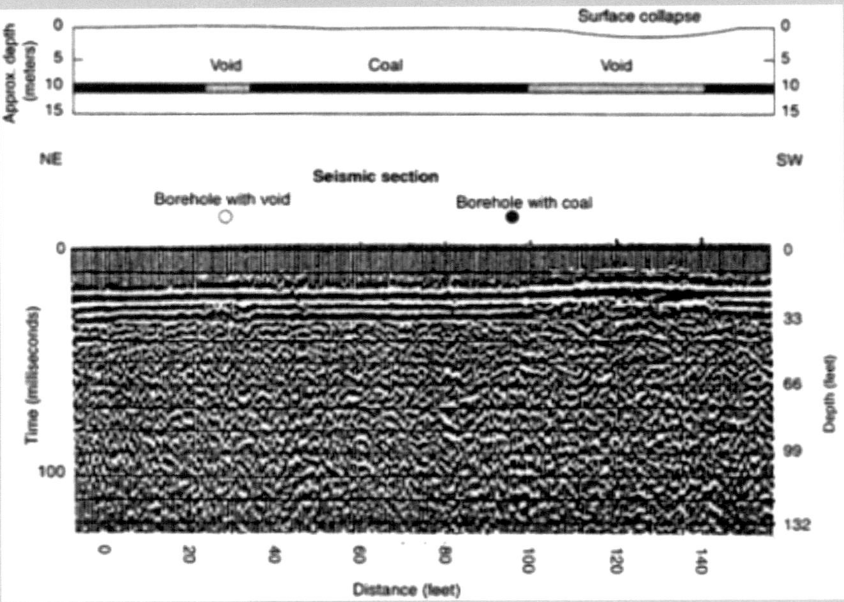

Seismic detection of mine void at 33-foot depth. From Branham and Steeples, 1988. Courtesy of Society for Mining, Metallurgy, and Exploration, Inc.

cone of influence of the injected fluid; this could be used to discern indirectly the coal outcrop barrier or seam's width. It should be noted that the interpretation of these data is complicated because the coal remaining to form the outcrop barrier or pillars would be under greater loading stress than the unmined coal seam due to the increased overburden pressure distributed on a in the expansion of the cleat void spaces, effectively increasing the hydraulic characteristics of the coal. Therefore, additional data for coal under high lithostatic pressure should be collected and evaluated for feasibility and development of this method. Research should be performed to ascertain whether hydraulic testing has merit as a cost-effective and accurate method to aid in determining the extent of coal outcrop barriers and coal voids in mines adjacent to coal waste impoundments.

SUMMARY

One of the critical tasks in site characterization is ruling out the presence of voids. Invasive drilling programs can provide the necessary information. However, they may compromise the hydrological integrity above the mine, and their cost is often significant, both economically and environmentally. Well-planned and appropriate use of geophysical techniques can often help to minimize the amount of drilling required to detect mine voids. However, no single geophysical technique will work at all depths in all types of geology. From a practical standpoint, steep topography compounds the difficulty in collecting, processing, and interpreting geophysical data when surface methods are used, but these effects are minimized when borehole, cross-hole, and in-seam methods are used. In addition, trees and cultural features such as fences can impede geophysical data collection, processing, and interpretation. Multiple geophysical techniques may be necessary to reduce the probability for error to an acceptable level; drilling is required for confirmation.

The committee concludes that geophysical techniques are useful in some cases in coal mine void detection, especially the use of seismic surface waves, seismic reflection, ground-penetrating radar, and electrical resistivity methods. The committee also concludes that geophysical techniques have been underutilized in the coal-mining industry and could benefit from additional research. **The committee recommends that demonstration projects using modern geophysical techniques be funded, and that the results be widely conveyed to the mining industry and to government regulatory personnel through workshops and continuing education.** Continuing education could include the opportunity to attend short courses and seminars that present the latest technology along with case histories to support its use.

The committee notes that much more work has been done using geophysical techniques on coal field problems than is indicated in the literature. Since a large amount of the work is proprietary or involved in litigation, little has been published. The committee notes that publication of case histories on this work would be desirable.

6

Limiting Potential Failures

A coal waste impoundment can be viewed as a system composed of a series of elements that include the embankment, principal and emergency spillways, runoff diversion structures, the basin, barriers, slurry delivery and water recovery systems, and any additional drainage systems installed to protect the integrity of the structure. Analysis of the ways individual elements can fail when coupled with analysis of the impacts a specific failure or set of failures will have on the entire system allows a designer to evaluate ways to mitigate effects of individual failures. The objective is to design and assess facilities so as to avoid failures that compromise the impoundment system integrity.

An embankment or basin can each fail in a number of ways. It is essential to understand these failure modes and take appropriate measures to mitigate them. This chapter examines embankment and basin failure modes and mitigative measures. Particular emphasis is placed on basin failure modes and mitigation activities, because the largest remaining uncertainties for impoundments lie in the characterization of the basin area and in the mitigation of risks associated with the breakthrough potential (see Chapter 3). Risk assessment of new and existing impoundments is the first step toward risk reduction. Once the impoundments with the highest risk are identified, various methods can be explored to manage or reduce the level of risk.

Failure can be initiated through faulty construction and operation of coal waste impoundments. The role of the impoundment operators is discussed, including the establishment of best practices and management systems and emergency planning and risk communication.

Creating an impoundment, particularly near the head of a valley, which is usually a groundwater discharge zone, can also have significant consequences for the local hydrogeological regime. In the worst case, changes in the local hydrogeologic flow setting caused by an impoundment, could contribute to mine-associated blow-outs that can cause flooding or other environmental damage. Although a detailed analysis and discussion of blow-outs is not within this committee's Statement of Task, the hydrogeological implications for impoundment siting should not be neglected in design.

A comprehensive systems approach should be applied to the entire impoundment structure when analyzing potential failure modes. While it is important to view the entire impoundment as a system, it is also useful to review the modes of failure that can occur for major components of an impoundment—the embankment and the basin. The next sections provide that review.

EMBANKMENT FAILURE MODES

An embankment to contain coal mining waste is similar to an embankment to store and contain mine waste in other extractive industries, where they are usually called tailings dams. While the nomenclature is different in the coal business—refuse impoundment versus tailings dam, fine refuse versus tailings, and coarse refuse versus waste rock—the concept of impounding slurry behind an engineered embankment is the same. Coal refuse impoundments are a subset of a larger group of mine slurry impoundment systems called tailings dams.

It is common for tailings dams to be of the upstream, centerline, or downstream types (see Chapter 3). While the design and construction of tailings dams draws on the technology used for water-storage dams, tailings dams differ from water-storage dams in a number of important ways. They store primarily mine waste and only secondarily water. Tailings dams are often constructed from components of the mine waste stream and are usually built by mine operators over the life of the mine. Finally, the allowable seepage may be more restricted than with water storage because of environmental concerns.

There have been a number of worldwide surveys of tailings dam failures (e.g., ICOLD, 2001; Martin and Davies, 2000; USCOLD, 1994; Vick, 2000). Because the range of experience in these reviews is much greater than that in coal waste embankments alone, they provide a record of potential failure modes, many of which are directly relevant to failure modes that could befall coal waste embankments.

The U.S. Committee on Large Dams (USCOLD, 1994) defines incidents as dam breaks or loss of impoundments leading to a release of tailings and impoundment fluids; accidents that stressed the dam in some form without release; and groundwater contamination. A primary factor differentiating incident cause was the type of dam construction. The distinction of causes in terms of their relative proportion differs between upstream-type and downstream-type dams (S.Vick, consultant, personal communication, 2001). The preponderance of failures occurred during active operation of the impoundment with surface water on the tailings deposit. By contrast, only a few failures occurred for inactive impoundments subsequent to water removal upon abandonment. These were caused mostly by overtopping attributable to inadequate post-closure spillways (S.Vick, consultant, personal communication, 2001). Slope instability and earthquake effects dominate failure causes for upstream-type dams. However, seepage, overtopping, and foundation instability are other important considerations.

In recognition of the need to improve the design and construction of tailings dams, a number of guides to good practice have been prepared (e.g., ICOLD, 1989b, 1994, 1995a, b, 1996). Frequency of failure has declined during the past decade, which has been attributed to more rigorous engineering and regulations (Cambridge, 2001).

The failure at Buffalo Creek (Sidebar 1.3) was a pivotal experience in U.S. practice with regard to the need to improve design and regulation of coal waste embankments. Since MSHA was established, there have been no incidents of embankment instability, other than overtopping of starter dams early in construction. Nevertheless, it is prudent not to be complacent.

Worldwide experience with upstream-constructed tailings dams indicates that many, particularly those with wide subaerial beaches, have performed well in significant seismic events, when subjected to intense rainfall, and sometimes in spite of questionable operating practices. Davies and Martin (2000) outline the requirements for construction of a safe upstream tailings dam. Unless alternative processing and disposal methods are adopted on a large scale (Chapter 7), embankments in the future will likely be higher than in the past. *Given this challenge, and given the fact that some modern dams in other extractive industries have failed, the committee concludes that it is essential that MSHA and OSM stay current by ensuring that design criteria reflect the latest experience from all segments of the mining industry.* Although the committee has not identified any deficiencies, it is a matter of due diligence that MSHA, OSM, and industry employ the best available current technology. **The committee recommends that MSHA and OSM continue to adopt and promote the best available technology and practices with regard to the site evaluation, design, construction, and operation of impoundments.** For example, MSHA and OSM should commission periodic reviews of existing technical procedures and practices, with particular attention to the basin. Results of the reviews should be disseminated to industry. Based on the outcome, MSHA and OSM may have to revise guidelines to establish minimum expectations and levels of investigation for site characterization, design, construction, operation, and closure of coal refuse impoundments.

Structural Stability

Structural stability is based on the same general principles as the stability of water-storage dams, and structural failure generally occurs by the same processes: slope instability, liquefaction (commonly due to seismic activity), and foundation failure. Evaluation of slope stability relates the resisting forces of the embankment and its materials to the driving forces of the impoundment. Evaluation of the seismic stability of the embankment is based on the seismicity of the site and the potential for the embankment material to liquefy or lose strength during shaking.

The committee has identified the following factors that merit special attention in stability assessment, in particular for upstream-constructed embankments:

- *Long-term durability of the coarse refuse.* Since the coarse refuse is a component of the embankment structure, its resistance to weathering or durability should be assessed with respect to future settlement and shear strength. As discussed in Johnson (1999) and Linsey et al. (1982), the long-term performance of rock depends on the rock type as well as the site climate and setting for which the rock is used. Differing rock types vary in durability, with certain sedimentary rocks more susceptible to weathering due to the presence of clay minerals, which expand with moisture. Where coarse refuse is used for critical portions of the embankment, its durability should be assessed for the presence of these weathering minerals.
- *Shear strength and higher embankments.* For rockfill embankments, the shear strength characteristics of the granular materials in the embankment may change with increasing stresses as individual particles crush or break (Leps, 1970; Marsal, 1973; Wilson and Marsal, 1979). The shear strength and performance of embankment materials for a smaller embankment, for example, may not be the same as those for a high embankment because of particle crushing. The performance of the coarse refuse should be confirmed by shear strength testing under anticipated loading conditions.
- *Increasing clay content in fine refuse.* Fine refuse impoundments contain increasing amounts of clay, because more coal seams are

being mined that contain increasing amounts of clay partings. Since the clay particles typically slow the rate of consolidation and reduce shear strength properties of mill tailings and fine refuse (Vick, 1990), the consolidation and shear strength characteristics of the fine refuse should include the anticipated amount of clay that would be present.

- *Seismic evaluation.* Coal refuse impoundments are designed for stability under anticipated static and seismic conditions during the critical phases of construction and operation. For closed refuse impoundments, these stability conditions still apply, but the period for acceptable performance is indefinite. This means that for evaluation of seismic conditions (as well as precipitation events), events representing long-term recurrence intervals are required (such as 500-year to maximum credible earthquake loading conditions). These long-term seismic loading conditions, coupled with whether the refuse impoundment remains saturated, should be included in the evaluation of the post-closure performance of slurry impoundments.

Seepage and Piping

Seepage through embankments can lead to failure by internal erosion, the process commonly known as piping. In design and construction of water-storage embankments, the possibility of seepage-caused piping is commonly prevented by installation of filters or drains. Although the use of internal filters or drains in embankments is not common practice in Appalachia, drainage through the embankment is an important consideration in the design and construction of the structure. The coarse coal refuse filters the fines while allowing clear seepage to flow. The drains that are included provide a secondary line of defense for the control of seepage. The lack of an internal filter is acceptable, but it places an extra burden on high-quality compaction control and the use of subaerial beaching at various locations around the perimeter of the impoundment (the beach area created near the discharge point of the slurry transfer pipe) as an additional line of defense. Beaching should also be considered as a line of defense around the basin.

Overtopping

A coal refuse impoundment must be designed so that inflow does not exceed storage and outflow capacity. If inflow exceeds storage plus designed outflow (including that handled by spillways, decant facilities, diversion structures, and evaporation), the dam will be overtopped. Overtopping can cause substantial erosion of the embankment crest, which, if left uncontrolled, will usually work progressively downward, releasing water and coal refuse downstream.

Inflow includes: direct precipitation onto the impoundment area; runoff from the contributing drainage basin due to precipitation or snowmelt; groundwater inflows to the basin; outflow from other ponds in the basin; mine-water disposal; and preparation-plant slurry. Especially important are natural floods produced by major storms. The designed precipitation event used for a coal refuse impoundment will vary depending on the hazard classification of the facility, but in nearly all cases in the Appalachian region will be the probable maximum precipitation event. This criterion is conservative in that designers are obliged to provide sufficient storage in the facility to contain direct precipitation from the probable maximum precipitation event plus all other influent fluids from processing and runoff and still maintain 3 feet of freeboard. In determining the probable maximum precipitation event, designers rely on precipitation records and storm recurrence intervals to predict severe storm events such as hurricanes, as well as the effect of a reasonable foreseeable rain or snow runoff event. As discussed in Chapter 3, diligent slurry and water management is critically important to an effective coal waste impoundment system.

BASIN FAILURE MODES

Basin integrity is a routine consideration in all impoundment design. However, mining near a basin introduces special problems. In the evaluation of the basin of an impounding structure, the hydrogeological parameters are key in the determination of potential failure. The control and understanding of the leakage pathways—such as subsidence, excessive seepage, or internal erosion—are essential to determine the stability of the impoundment basin and the effect on the water balance and to comply with regulatory issues of these discharges. These are common considerations in all impoundment designs. It is essential that attention be paid to the identification, evaluation, and mitigation of potential failure modes in the basin.

Subsidence

Subsidence disturbs the strata above and adjoining the mined area and is an important failure mode (see Chapter 3). It implies opening of tensile cracks on the surface, displacement along faults and joints, separation along bedding planes, and some distortion of the strata around the workings. The immediate roof tends to cave into the workings, and the floor may heave. Subsidence movements may combine to create leakage pathways from the bottom of the impoundment to the outside environment. Site characterization, design, operation, maintenance, and monitoring require additional considerations in areas with subsidence potential. Chapters 4 and 5 review ways to detect such problems in the basin area.

Mine Openings

As noted in Chapter 3, the position of underground mine workings relative to an impoundment is a factor in the evaluation of the potential for breakthrough. The relative elevations of local drainage and slurry height are also critical.

If unidentified or not properly sealed, mine openings act as conduits for the flow of slurry or "black water," which may contaminate waterways outside the mine. Even if the workings do not allow water to flow out, there may be connections to local aquifers or other permeable strata. This may lead to seepage far from the impoundment. Therefore, identifying mine openings in or adjacent to the basin is of paramount importance.

Basin failure above active or abandoned mine workings may involve one or more of the following modes:

- *Leakage along naturally occurring joints and fractures.* Joints and fractures may fill with ultra-fine material but do not necessarily develop resistance to the flow of fluids. Cleats in seams make the coal more permeable and are included in this category. The August 1996 failure at Lone Mountain was apparently caused by a breach in the mountainside (K.Mohn, Lone Mountain Processing, personal communication, 2001).
- *Subsidence-induced tension or shear cracks and fractures at the bottom of the impoundment pool.* Slurry tends to lubricate the joint and fracture surfaces and may promote movement, rather than inhibit flow. Entry of water between strata along bedding planes could accentuate subsidence effects and result in failure. The

October 1996 incident at Lone Mountain appears to have been induced by subsidence (K.Mohn, Lone Mountain Processing, personal communication, 2001).

- *Sinkhole or pit subsidence.* If the interburden between the pool floor and the room-and-pillar workings is shallow (100 feet or less), sinkhole or pit subsidence may occur. This could cause a sudden inflow of the impoundment material into the mine openings. The November 1996 breakthrough into abandoned mine workings at Buchanan Mine was of such origin (B.Thacker, Geo/Environmental Associates, personal communication, 2001).
- *Catastrophic events.* Heavy rainfall, debris flows, sudden snowmelt, breakage of seals, or blow-outs in mines above the slurry level would impose a severe load on the basin floor and could cause collapse of the pillars in workings below, creating a connection. The failure in Martin County in October 2000 occurred after heavy rain fell in a short period.
- *Piping.* Piping due to erosion may occur once seepage or leakage has been established.

If the width of the outcrop coal barrier or the overburden above the mined area are of insufficient thickness or the surrounding strata are too weak, impounded water could break into a mine and lead to a basin failure (see Chapter 3). Conversely, an inundated mine under higher hydraulic head could introduce a large volume of water into the impoundment. Should a blow-out occur elsewhere in the watershed above pool level, it could cause an influx of substantial amounts of water into the impoundment. This, in turn, may overtop the embankment, damage the spillways, or induce a breakthrough in the pool bottom as a result of the additional load. Synergy between geologic and hydrogeologic conditions can compound the instabilities created by any failure mode (Figure 6.1).

Currently, no federal regulations address the width of outcrop barrier that should be left during underground coal mining. Kentucky and Virginia currently have standards for outcrop barriers, but allow variances where conditions are appropriate. In some cases it is prudent to allow openings for drainage. OSM has studied the problem of outcrop barriers but has yet to release any conclusions to date. **The committee recommends that MSHA and OSM jointly pursue the issue of outcrop coal barrier width and overburden thickness and its competence and develop minimum standards for them.**

Mine workings below an impoundment should be avoided unless they can be confirmed to be deep enough and to contain an aquitard layer

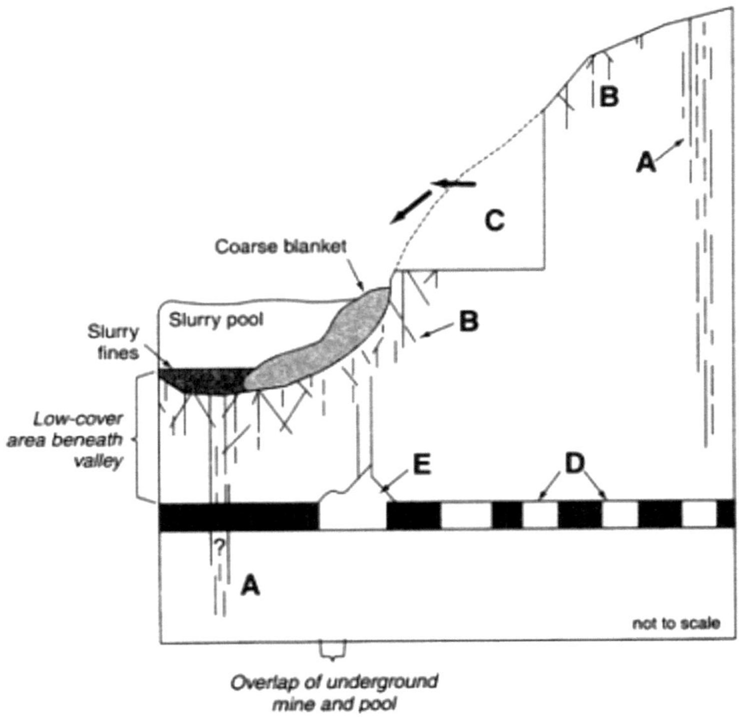

FIGURE 6.1 Cross section of a hypothetical waste impoundment and geologic features that control site stability. (A) Deeply penetrating tectonic fracture sets. (B) Shallow, near-surface or stress-relief fracture zone. (C) Rock cut from contour to supply a coarse blanket for bank stabilization and support. (D) Mined out areas of an underground coal seam. (E) Roof collapse above mine void, creating a zone of structural weakness. Modified from S.Greb, J.Dinger, and D.Cumbie, Kentucky Geological Survey, personal communication, 2001.

adequate to prevent uncontrolled entry of water into the openings. The strata in the aquitard must be identified and their thickness determined. This must be done with site-specific investigations and rock-properties data. If specific data cannot be obtained where longwall or other full extraction pillar recovery has occurred, the surface fractured zone may be considered to be from 50 to 200 feet thick (Kipp and Dinger, 1987; Singh and Kendorski, 1981). The permeable zone immediately above the openings may be 60 times the thickness of the extracted seam, if sufficient width of opening with respect to seam depth has been excavated. It should be recognized, however, that these values may have to be modified for known site conditions.

MITIGATIVE MEASURES

The ability to intervene in the event of undesirable performance and to introduce mitigation measures is an integral aspect of eliminating basin failure. The ease of doing so depends upon whether slurry elevation has already exceeded the level of the above-drainage coal seams or associated workings.

If the basin is new or if the slurry elevation has not yet exceeded the level of above-drainage coal seams, the quality of the existing outcrop barrier between the basin and the coal seam should be evaluated. If the outcrop barrier is inadequate, it should be enhanced. A number of enhancement measures are available, including construction of additional barriers, installation of drains, promotion of seals such as what might occur with the consolidation of the slurry itself, and reinforcement of the existing barrier by grouting or other means. The solids deposited at the bottom of the impoundment can, under some circumstances, consolidate enough to create an impervious blanket of cohesive material. This reliance on the impervious zone can be evaluated using well-established procedures in geotechnical engineering.

In the case where the slurry elevation has not exceeded the level of the above-drainage mine workings, the assessment of the outcrop barrier is generally feasible. However, if the condition of the outcrop barrier cannot be adequately assessed, the risk associated with developing a new facility or continuing with a preexisting one increases. If the costs of mitigation or the risks associated with the proposed operations are excessive, an alternative disposal strategy, which might even entail an alternate site, becomes necessary.

When slurry elevation exceeds the level of the above-drainage workings, it becomes more difficult to assess the quality of the existing barrier. This may result in increased cost of investigation or greater uncertainty with the results or both. For example, as discussed in Chapter 5, the use of geophysical and remote sensing techniques in and adjacent to the basin area may reveal anomalies related to voids or geologic features such as fractures. These anomalies can provide clues about the potential mechanical and hydrogeological conditions in the proposed impoundment area. In the absence of geophysical, remote sensing, or hydrogeological anomalies, clearing of the basin area may be warranted to uncover geologic anomalies that might otherwise go undetected. While the steps to be taken in the selection of mitigative measures are the same as noted above, many of the measures become more difficult to implement (Sidebar 6.1). Hence reliability associated with assessment and intervention in this case is reduced. This can be overcome, to some degree, by increased reliance on monitoring.

The assessment of below-drainage workings constitutes a subset of the second case above. Here the integrity of the cover must be evaluated by some combination of drilling, geophysics, fracture analysis, and subsidence assessment. A distinction should be made between active and inactive workings, because the safety of miners is at stake in active workings. Here, if basin sealing cannot be relied upon, and alternative methods of slurry management cannot be practiced, as may be in steep topography, then the size of the impoundment may have to be restricted.

Coal seams above the maximum permitted slurry elevation may also be susceptible to blow-out. In this case, the assessment requires an understanding of the hydrogeological circumstances and whether they are favorable for blow-out. If so, mitigation is appropriate and some combination of drainage, sealing, and barrier construction, together with monitoring, is needed.

The committee concludes that selecting the appropriate mitigative measures relies strongly on reliable basin characterization. **The committee recommends that MSHA and OSM develop and promulgate guidelines for the site evaluation, design, construction, and operation of basins.** They should be comparable in scope to the guidelines used in embankment design.

If slurry from an impoundment leaks into active or abandoned mine workings, or may do so, bulkheads or seals may be constructed to preclude the water from escaping into the outside environment (Chekan, 1985). As discussed above, many mitigative measures can be designed using established procedures; bulkheads designed to support high hydrostatic pressure present a different kind of problem. **The committee recommends that MSHA review its current practice and develop guidelines for the design of bulkheads intended to withstand hydraulic heads associated with slurry impoundments.** The bulkhead should be constructed of material that can withstand water action without deterioration in the presence of the various chemicals in the impoundment water. Further, the bulkhead should be suitably anchored in competent, unfractured strata. If such an area is not available, pressure grouting may be needed. Deterioration of the anchoring strata can be a major structural problem where the bulkhead is keyed into

SIDEBAR 6.1 MITIGATION MEASURES AT LONE MOUNTAIN AND BUCHANAN, VIRGINIA

In August 1996, at the Lone Mountain, Virginia, slurry breached the impoundment and went into abandoned mine workings that were much closer to the pool than was shown on mine maps. The area was excavated to expose underlying bedrock. The exposed workings were sealed by backfilling with competent rock. This was covered with a geo-textile and further backfilled with compacted earthen materials. Backfill was placed over the seal to an additional depth of 20 feet and compacted. To protect the area further, along the western wall of the impoundment pool, a barrier of compacted earthen materials was constructed. In October 1996, two large sinkholes developed at a different location. Excavation revealed a fracture in the roof of old workings that had allowed slurry to enter the mine. The loose rock was removed, and the void was sealed with polyurethane grout. The exposed rock face was covered with a geo-textile and backfilled with compacted earthen materials (K.Mohn, Lone Mountain Processing, Inc., personal communication, 2001).

In November 1996, the impoundment at Buchanan, Virginia, failed and slurry entered old workings created by a different mining company. The area was excavated, exposing filled auger holes. Mine workings were backfilled; a filter fabric was placed along the entire perimeter of the coal seam; cohesive soil fill was compacted in lifts to create a barrier, and coarse refuse was then used to backfill the remaining excavation; and French drains were installed around the entire perimeter of the facility to drain the coarse refuse perimeter embankment (B.Thacker, Geo/Environmental Associates, personal communication, 2001).

water-sensitive, clay-bearing strata. The size, integrity, and strength of the surrounding coal pillars, roof, and floor are critical to successful sealing. Generally, seals constructed for ventilation cannot withstand the anticipated water pressures.

IMPOUNDMENT MANAGEMENT

Risk reduction cannot be achieved by design and regulation alone, but requires adopting the best available construction and operating practices. The annual inspection review provides one check in this system, but it may not identify all construction and operation problems. Experience with tailings dam failures suggests that all concerned with safe impoundment management must pay additional attention. The International Committee on Large Dams (ICOLD, 2001) recently summarized lessons learned from tailings dam failures. Its recommendations include: detailed site investigation; state-of-the-art procedures for design, construction, and operation; routine monitoring; safety audits; and occasional specialist reviews.

Experience elsewhere indicates the desirability of corporations developing an impoundment or tailings management system that addresses policy, commitment, planning, implementation, checking, corrective action, and management review. The guidelines developed by the Mining Association of Canada (MAC, 1998) are an example. Coal operators are integral stakeholders in risk management and reduction. The committee suggests that coal operators develop an industry-wide procedure for evaluating impoundment management systems that could be adapted to specific properties and corporations.

RISK ASSESSMENT AND MANAGEMENT

In response to the basin failure at Inez, Kentucky (Sidebar 1.12), MSHA and a number of state agencies have initiated surveys to assess the risk associated with current impoundments. The committee agrees that identification of those impoundments within the existing inventory that have the greatest risk of failure has significant value. However, the two classification systems MSHA currently uses (see Chapter 1) are not consistent with accepted definitions of risk. The risk associated with a failure is defined as the product of hazard (the potential for a failure to occur) and the consequences of that failure (loss of life, costs of repairing damage to structures or facilities, environmental impacts). An impoundment could have relatively low risk if consequences of a failure were low even though the probability of a failure was moderate. On the other hand, an impoundment with a low probability of failure could be assigned high risk if the consequences of failure, in terms of danger to human life, damage to valuable structures or the environment were large. The pair of MSHA classification systems currently use the term hazard in a way that is not consistent between the two classifications system. In the first classification scheme, which deals with potential impacts of embankment failures, the term hazard refers to the consequences of failure. That classification system makes no attempt to assess the probability of a failure event for individual embankments. The second classification scheme, which deals with basin failures, comes closer to the standard definition of a risk assessment. That system includes an assessment of the proximity of underground workings and the potential for a failure that would lead to a release of water or slurry from the basin area into underground mine workings. In addition, the potential impacts of such a failure are considered. Thus, the second classification scheme is a version of a risk assessment, according to the standard definition. *The committee concludes that using different hazard classification methodologies for embankments and basins is inappropriate.* **Therefore, the committee recommends that: (1) MSHA and OSM review activities related to risk assessment for existing impoundments (including both embankments and basins) to ensure that they are consistent and that they distinguish appropriately between hazard and consequence assessment in the methodologies adopted; and (2) MSHA and OSM establish a single, consistent system, which should be used to assign both embankments and basins to risk categories.** The ranking should be based on the combination of hazards and consequences, such as loss of life, cost, and environmental impact. Proposed new impoundments should also be assigned to risk categories, based on a combination of hazards and consequences, as was suggested for existing impoundments. The committee believes that this can be accomplished using qualitative risk assessment techniques.

A consistent risk assessment system would allow decisions on impoundments to be based on their relative risks. **The committee also recommends that MSHA and OSM oversee a thorough assessment of potential mitigation measures for those impoundments that fall in the highest risk category and should determine which mitigation measures should be applied to reduce this risk to an acceptable level.**

Geotechnical engineering is used in the design and construction of waste impoundments. Given the inevitable uncertainties in site characterization, knowledge of material properties, and the need for use of idealized models to describe both physical and human behavior, risk is inherent. Therefore, managing risk is an essential consideration (Morgenstern, 1995).

Fortunately, powerful methods of risk management have evolved. The observational method is the first line of defense in managing risk in the face of identified uncertainties. This method involves the use of observation to review performance and refine subsequent design, construction, or operation. Peck (1969) identified the elements of the observational method as follows:

- Site exploration to establish (at a minimum) the general nature, pattern, and properties of subsurface materials;
- Assessment of the most probable conditions and most unfavorable deviations from these conditions;
- Establishment of the design based on a working hypothesis of anticipated behavior under the most probable conditions;

- Selection of quantities to be observed as construction (or operations) proceeds, and calculation of anticipated values of these quantities on the basis of the working hypothesis;
- Calculation of values of the same quantities under the most unfavorable conditions compatible with available subsurface conditions;
- Selection in advance of a course of action or modification of design of every significant deviation from that predicted for the working hypothesis;
- Measurement of the quantities to be observed and evaluation of actual conditions;
- Modification of design to suit actual conditions.

Because geotechnical uncertainty is intrinsic, designers, owners, and regulators should make conscious use of the observational method and accept the need to declare performance indicators, response procedures, and observational procedures, including advances in monitoring technology. The observational method of risk management is applicable to both existing and future impoundment systems.

The value of the observational method is well recognized in geotechnical practice, and MSHA regulations and current practice promote elements of it. However, the method has some limitations, such as difficulties in application to seismic or other rapid events. In particular, it requires anticipation of all eventualities as well as preparation for courses of action to meet whatever situation develops. MSHA and coal industry designers should be aware of the need to employ the observational method to the degree practical.

Another method of risk management is the use of third party reviews. The committee recognizes the value of third-party reviews that have been used in projects such as the design and construction of large water dams. Such reviews often examine whether the project is being designed to appropriate standards, and the construction is being managed appropriately, and the committee suggests that coal companies consider whether similar reviews would add value and help manage risk in the design, construction, and operation of coal waste impoundments.

To maximize the potential for risk reduction, the committee recommends that all impoundment designs be accompanied by a risk analysis utilizing qualitative methods. Examples of such methods include Potential Problem Analysis and Failure Modes and Effects Analysis. The performance requirements needed to correct failure modes, including the instrumentation to be installed to evaluate performance must be identified, and the plan to be executed in the event that performance is not met should be spelled out.

The committee believes there is a limit to risk tolerance, for both existing and new impoundments. When risk is high, and when mitigation, either through more reliable characterization or barrier constructing is impossible, of limited precedent, or so expensive that it is infeasible, then a substantial change in operation of the impoundment is warranted. This may range from minimizing slurry fluidity to ceasing operations. If an impoundment fails risk-assessment criteria and if risk cannot be mitigated it should be phased out or alternatives considered.

MONITORING

Monitoring is an integral part of the observational method used in geotechnical engineering. In the design and construction of coal waste impoundments, monitoring is critical since the construction process continues for the life of the facility. During the life of the impoundment, which may span decades or more, conditions may change. For example, the nature and characteristics of the refuse may differ because of the areal variations in the geology of the coal seam, mining of different coal seams, alterations in mining or preparation methods, or rate of waste generation. A well-planned monitoring program can help to detect when major changes are occurring so that design modifications can be implemented in a timely manner. The savings that may be realized by not designing for the most conservative scenario can justify the scale of the monitoring program. Monitoring can help ascertain when repairs, improvements, or other upkeep is needed.

Monitoring procedures and instrumentation are well documented (ASCE, 1999; Dunnicliff, 1988) and need not be repeated here. Monitoring instrumentation requires that appropriate target sites be identified and accessible. Monitoring of potential failure modes of embankments typically measures pore pressures, surface and internal deformations, hydraulic parameters, and vibrations, especially if blasting is being conducted nearby. Occasionally, temperature and rock stress or soil pressure, especially if high horizontal stresses exist in the area, should be measured. Hydrogeologic monitoring and downstream flow and quality measurements may give evidence that would provide warning of impending basin failures.

It cannot be overemphasized that instrumentation should be used as a complement to visual observation and not as a substitute. Instruments cannot adequately establish the extent of vegetation and undergrowth or its removal, the origination of new seeps, initiation of erosion, and many other parameters.

As discussed in Chapter 2, OSM and its state delegate programs require mine operators to monitor both surface and groundwater at coal mining sites. The vast majority of the work published on slurry impoundment composition and chemistry was conducted in the mid 1970s and early 1980s. However, considering the advances in coal cleaning techniques and in storage and abandonment of the refuse, and the likely impact on impoundment chemistry resulting from the implementation of these technologies, this data may not be representative of current conditions (Darrell Taulbee, Center for Applied Energy Research, Lexington, KY, personal communication, 2001). In addition, a theme mentioned repeatedly in town meetings with coalfield citizens was their concern and desire for information concerning the chemical constituents in the coal waste, and how it affects their ground and surface water. **As a result, the committee recommends that research be performed to identify the chemical constituents contained in the liquid and solid fractions of coal waste, and to characterize the hydrogeologic conditions around impoundments.** An additional benefit to this research is that the characterization of the wastewater will aid in monitoring schemes that could aid in leak detection, which could foreshadow impoundment failure (Sidebar 6.2). These monitoring systems focus on forensic petrologic and geochemical investigations that may aid in the early detection of fugitive solutions from a refuse impoundment system. While these investigations and monitoring programs are not directly germane to limiting the potential for refuse impoundment breakthroughs and failures, they may act as an early warning system for mine operators to change their management of the impoundment system. Conjunctive research should also be conducted to demonstrate the use of continuous data loggers and real-time monitoring devices that can monitor and warn of changes in hydrologic, hydrogeologic, and geotechnical conditions around the impoundments.

It must be clearly understood that for monitoring to be successful it should be applied to all modes of potential failure. Monitoring expenses can only be justified if data and analysis results are received in a timely manner. Acquisition and analysis of data should be sufficiently rapid to enable coal operators to make meaningful decisions. If data indicating movement or water pressure increases are not analyzed prior to a black-water release or failure, they are of limited value. Computerized processing can immediately transform the raw data into a format the engineer can use. If some crucial criteria are exceeded, visual or audible alarms can be triggered.

SIDEBAR 6.2 FORENSIC HYDROLOGY

Systematic changes in water chemistry are often used to detect plumes of contaminants around leaking landfills, underground storage tanks, and mine waste. Similarly, changes in the chemical character of water samples from ambient levels may indicate leakage from impoundments into the surrounding groundwater system. Hydrochemical facies are defined as areas of aquifers where the chemistry of the groundwater is predictable within defined limits. A hydrochemical facies model for the Appalachian coal field (Pennsylvania Department of Environmental Protection, 1999; Wunsch, 1993) can be used to detect groundwater contamination from coal waste. The water fraction of coal slurry will chemically reflect the concentrated waste material in coal and will also be the most mobile fraction to leak into the underlying or adjacent bedrock, coal seams, and fractures. For example, site-specific data show that coal waste effluent can be enriched in iron, aluminum, magnesium, and sulfate (D.Taulbee, University of Kentucky, personal communication, 2001). Moreover, organic chemicals used in the beneficiation of coal waste may also be used as a groundwater tracer to identify leakage from impoundments. A more comprehensive suite of analyses could be used to monitor and detect leakage from an impoundment by employing geochemical modeling, which can uniquely identify and characterize water samples. Thus, monitoring the chemical composition of water adjacent to impoundments could be used to detect whether water is leaking from the coal slurry and, with the other site-specific information, to determine sensible mitigative programs.

Several water quality parameters normally associated with mining impacts can be easily determined in-situ. For example, specific electrical conductance, which is a measurement of capacity of a fluid to transmit an electrical current, is directly proportional to increased solids or salt content in the water. Water from waste impoundments or mine waste usually contains increased dissolved salts, such that significant changes in electric current can be a predictive tool for leakage. The pH of mine waste-impacted water is often notably acidic, thus changes in pH can be indicative of contaminated ground or surface water leakage.

Digital water quality monitoring equipment (i.e., data loggers) can simultaneously record changes in several parameters that indicate water impacted by mining or mining wastes. These tools can also be used to obtain real-time water-quality data that can be accessed remotely over the Internet or by radio or satellite transmission, and can be used to monitor for leakage that may presage an impending failure. These instruments are relatively inexpensive (approximately U.S. $5,000) and can be used for predictive monitoring of hydrologic and chemical parameters that would warn of impoundment failure.

In conjunction with chemical data, water levels could be monitored to establish the hydrostatic head conditions around the impoundment. Digital data collectors or data loggers can be used to monitor remotely anomalous changes in hydraulic head in the adjacent strata or mine workings. This too may indicate changing head conditions between the slurry impoundment and surrounding aquifers or mines. Digital data loggers to monitor these changes can be equipped with alarms that give real-time warnings that head levels have surpassed a pre-determined threshold. The instruments are proven in the field, relatively inexpensive, and widely available.

Continuous monitoring could provide timely warning in case of impending failure of an embankment or basin. Its use with weirs, for example, is well established, and it can be used with other types of instrumentation. **The committee recommends that MSHA and OSM consider requiring additional continuous monitoring in specific instances and evaluate automation of monitoring instrumentation.**

EMERGENCY PLANNING AND RISK COMMUNICATION

All coal companies that operate a coal refuse impoundment system are required to develop emergency response and evacuation plans that describe what may happen and what should be done to limit the damage of any reasonably foreseeable event. Because Appalachian refuse impoundment systems are usually positioned at the head of a stream, communities are often situated immediately downstream. A failure or breakthrough from refuse impoundment systems has the potential for significant adverse consequences for downstream communities, infrastructure, and the environment. Dam break analyses and evaluation of potential flooding resulting from impoundment failure are commonly undertaken in emergency preparedness planning.

Public concern regarding emergency response and evacuation plans was a recurring theme in public comments made to the committee. Some residents were unaware of emergency evacuation plans; others had seen evacuation plans but disagreed with the logic behind the evacuation routes and would not have used the plan in the event of an emergency. Conversely, coal industry and regulatory agency representatives confirmed that these plans are being developed and shared with the public through the various community contacts (e.g., local fire departments, police, health care providers).

The lack of realistic communication constitutes a fundamental barrier to the industry's ability to make stakeholders aware of the risk associated with coal refuse impoundment construction, operation, and closure and of steps taken to mitigate that risk. *Based on these stakeholders' input, the committee concludes that communication concerning coal refuse impoundment system risk and emergency response between the industry and the local communities could be improved substantially.* The committee suggests that the industry take steps through the appropriate emergency response agencies to address these problems.

7

Alternatives for Future Coal Waste Disposal

Coal system activities range from mining, to processing, to utilization, to disposal, and involve resources, transportation, and the environment (Figure 1.2). Redesigning systems like this to eliminate or reduce substantially waste streams is the first priority of industrial ecology (Sidebar 7.1). In the coal system, this can be accomplished by identifying the appropriate point in the process for reducing or removing noncombustible material. Currently, waste is created during the mining and processing stages and disposed of either in coarse refuse piles or fine waste slurry impoundments. Consideration of the entire coal system concept leads to the identification of alternatives to impoundment disposal of fine coal waste by reducing the amount of waste generated, utilizing waste, or disposing of waste elsewhere. These options attempt to minimize the waste stream, transfer it to another part of the coal system, or redirect it, respectively.

Fine coal waste from the preparation plant can be reduced or eliminated during both the mining and the preparation phases. For example, selective mining reduces the amount of noncombustible material. However, many of the coal seams currently mined are of sufficiently low quality that they would not be mined if this were required. Options for preparation plant waste elimination are more numerous during coal preparation. Cleaning requirements are usually imposed on the mine operator, whose coal cleaning and waste disposal may be constrained by site and environmental considerations. One option is to transfer cleaning responsibilities to the power plants (the majority of end-users). Another is to add fine coal waste to the cleaned coarse coal product. Dewatering the slurry product solidifies (to a degree) and reduces the volume of the waste.

Since 90 percent of the coal mined in the United States is used in power plants (Freme, 2000), significant benefits can be achieved if advances in power plant technologies are integrated into the coal system components. Emerging power plant technologies, developed around fluidized-bed combustion with appropriate flue gas desulfurization technology, allow burning of low value coals, i.e. coals with a large amount of noncombustible material (ash). Another technology is the use of coal water slurry as fuel for traditional combustors or gasifiers.

There are several alternatives to disposing of fine coal waste in impoundments, such as disposing of it in surface fills and underground workings. However, these options are often limited by factors such as topography, cost, and safety.

The ramifications of all potential options may be explored to create a Type III industrial ecology model for the coal/electric power industry (Sidebar 7.1). Figure 7.1 shows that through selective mining and other means, coal preparation needs can be restricted to Level 3 (Sidebar 7.2) and below, thus producing only limited fine coal waste slurry. If coal is cleaned to Level 4, the options of waste utilization and alternative disposal locations remain. Finally, existing ponds can be reclaimed, and the fine waste can be used in a similar manner.

SIDEBAR 7.1 INDUSTRIAL ECOLOGY

In the past, many industries operated as individual entities; however, the individual operations can have widespread impacts. The philosophies and approaches of industrial ecology may be used to integrate an individual industry, such as coal mining, with natural ecosystems and other industries. Industrial ecology is defined by Graedel and Allenby (1995) as a rational way for humans to maintain their existence with changing economies, cultures, and technological capabilities. Individual systems must work with each other to optimize the total materials cycle, including resources, energy, and capital.

In the biological world, metabolism is the key process for life, for ecological balance, and for providing an increased capacity for living things. When the ecological balance is disturbed, species perish or mutate until a new balance is established. Industrial metabolism adapts the concept to the industrial world. Related industries in the industrial system are designed to work together to imitate or mimic the metabolic process. This process does not exist in nature. Jelinski et al. (1992) present three models of industrial ecosystems based on this analogy (see figure next page). In Type I, the flow process is unidirectional from resources through consumption to waste. With time, the system's resources will be depleted, and its wastes will overwhelm it. In Type II, internal cycling loops are developed, leading to limited input of resources and limited waste. This system is also not sustainable because input flows to waste in only one direction. In the ideal Type III system, industrial processes are similar to the biological ecosystems model, and full cyclicity is achieved.

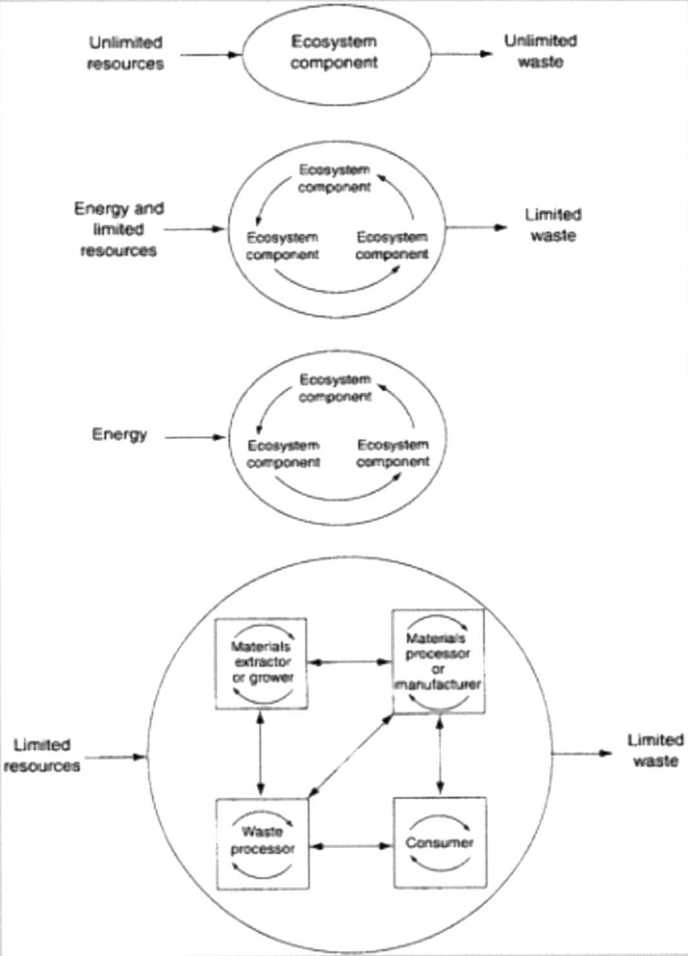

Reprinted with permission from Jelinski et al., 1992. Copyright 1992 by J.L.Jelinski.

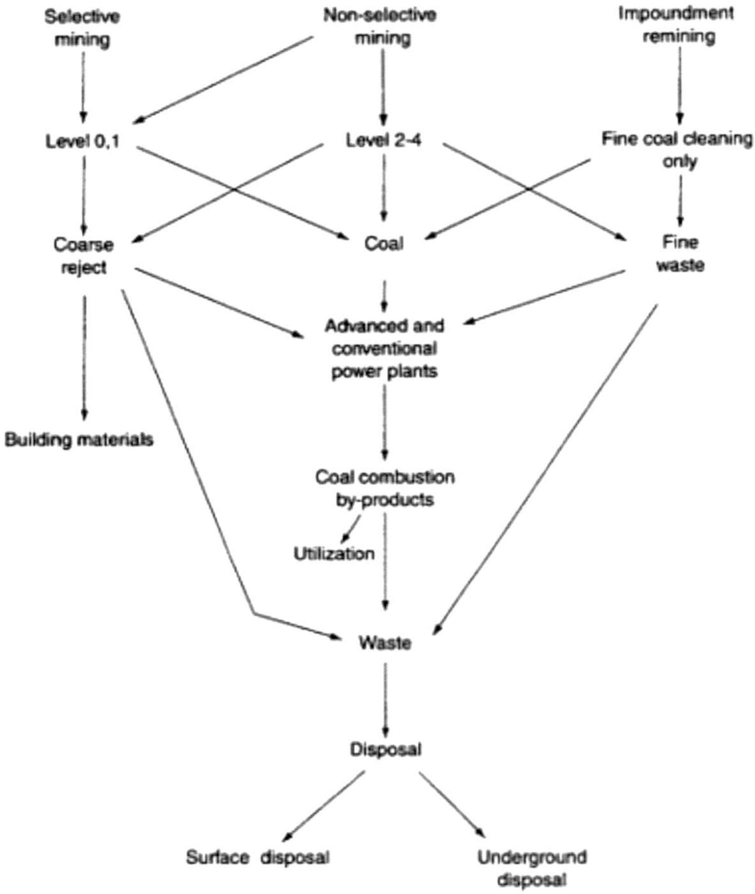

FIGURE 7.1 Coal system: Mining, preparation, utilization, residues, and disposal.

However, these options for impoundment alternatives raise other issues. For example, although the utilization of waste as a fuel eliminates the need for disposal of fine coal waste from the preparation plant, the burden of generating and disposing of coal combustion by-products is transferred to the power plant. Although this material has been utilized in a number of innovative ways (e.g., construction aggregates, synthetic soils, neutralizing agent), whether it is truly benign has been questioned. Another question is the safety and engineering aspects of alternative disposal locations, such as underground injection of slurry.

Proven technologies can abet some of the issues with slurry disposal. To make this a reality, major institutional, organizational, environmental, and business issues must be addressed to encourage a shift from the traditional mining, processing, and utilization practices to an approach based on industrial ecology. Several alternatives are outlined in the sections that follow.

REDUCING OR ELIMINATING SLURRY GENERATION

Of the more than 1 billion tons of coal mined in the United States, only about 350–400 million tons are cleaned in wet coal-processing circuits (C. Raleigh, CQ Inc., personal communication, 2001). The opportunities for reducing slurry volume include mining and coal processing alternatives.

Modern methods of surface and underground coal mining offer only a limited possibility for quality control during mining. Mining operations can be planned to extract coal from the best quality seams and minimize dilution with noncombustible material. This approach is commonly used in both surface and underground mining, especially for coal in the western United States. However, it is more difficult to apply in the eastern United States, where the highest quality seams have already been mined. Run-of-mine coal from both high and low quality sources can be blended to make a product of direct marketable quality.

When coal is cleaned in wet-processing circuits, a fine waste stream containing water, fine coal, and noncombustible particles (ash) is produced in which the percentage of each depends upon the level and efficiency of the fine coal cleaning methods employed (Sidebar 7.2). Slurry volume can be reduced by improving fine coal recovery and minimizing the mass of solids for disposal. The slurry volume can be further reduced by dewatering, which increases the proportion of solids to water. The ability to do either or both of these depends on the method of extraction, the amount of slurry dilution, the characteristics of the coal (e.g., the hardness of the coal, which affects the

SIDEBAR 7.2 LEVELS OF COAL PREPARATION

There are five levels of coal preparation.

Level 0—The coal is not cleaned, and run-of-mine coal is shipped directly to the customer.

Level 1—The run-of-mine coal is crushed to below a maximum size, and undesirable constituents are removed. The product of Level 1 preparation is commonly termed “raw coal.”

Level 2—The product from Level 1 is sized as coarse and fine coal. The coarse coal is cleaned to remove impurities; the fine material is added to the cleaned coarse coal or marketed as a separate product.

Level 3—The fine product from Level 2 is sized into two products: intermediate and fine. The intermediate product is cleaned to remove impurities. The fine material is added to the cleaned product or marketed separately.

Level 4—Cleaning is extended to the fine material from Level 3.

SOURCE: *BTU Magazine, 1982.*

size and amount of fine particles produced), and the local geology (e.g., abundant clay in adjacent strata can produce a refuse stream that is more difficult to dewater).

It has been nearly 20 years since dry coal preparation methods were used in the U.S. coal industry. According to Couch (1991), in the early 1960s, dry coal cleaning accounted for about 10 percent of all coal that was cleaned, but since then it has dropped to less than 1 percent. Dry cleaning is usually accomplished with “air jigs” or “air tables” using oscillating and fluidized bed principles. Most of the dry methods require closely sized and moisture-free feed. A combination of factors associated with the dry methods—particle size, dust, transport, health and safety, noise, and the better performance of wet processes—have contributed to the near abandonment of dry coal cleaning processes in recent years. The increased use of water for dust control in underground mines, and the increased efficiency of wet cleaning methods have continued the sharp decline in the use of dry cleaning methods at the mines. However, dry coal preparation methods, which do not create the same disposal challenges as slurry waste, can be effective in areas where the water supply is restricted.

Currently, most coal preparation plants recover fine coal only from the size fraction greater than 100 mesh (150 micrometer). Typically, this is done with water-only cyclones, spirals, or both for 16 × 100 mesh (1.0 × 150 micrometer) material. Particles smaller than 100 mesh (150 micrometer) account for 3 to 7 percent of the total plant feed, of which coal may comprise as much as 50 percent (R.Honaker, University of Kentucky, personal communication, 2001). These fine coal particles may be untreated, partially treated, or fully treated.

With no treatment, the fine material, usually smaller than 100 mesh (150 micrometer), is sent to a thickener and then pumped to a slurry impoundment. Alternatively, the fine coal and refuse could be blended with the coarse coal product and sent to the power plant. This feed can be either burned directly or further cleaned in the power plant after pulverization; advances in magnetic and electrostatic separation hold promise for dry cleaning the coal at this stage. In fact, pulverization liberates more of the mineral matter in coal. Dry cleaning with magnetic and electrostatic separators has shown encouraging results (Oder et al., forthcoming). However, this process still generates a waste stream (albeit not a slurry) that requires disposal.

Partial treatment utilizes classifying cyclones to remove coal particles to approximately 325 mesh (45-micrometer) size. In full treatment, which theoretically captures all of the fine coal particles, the fine waste from the cyclone is subjected to flotation. As air bubbles rise through the flotation tank, the coal particles attach themselves to the bubbles and are carried to the top of a column of water.

Although the recovery of all or most of the coal fines will not eliminate the need for slurry impoundments, it will reduce the required disposal volume. At the same time, by increasing the clay content, a slurry that is more difficult to stabilize may be produced.

In contrast, dewatering the refuse stream could eliminate the need for slurry impoundments by changing the strength properties of the waste material, although disposal of the resulting dewatered waste raises other issues. Dewatering employs either sedimentation or filtration or both. In sedimentation, the liquid is constrained, and the solid particles move freely. This results in clarification of the liquid and thickening of the remaining slurry. In filtration, a medium constrains the particles while the liquid flows through. This is accomplished by screening and centrifugation (Osborne, 1988).

In coal preparation plants, wet refuse is usually sent to a thickener (Figure 7.2). The underflow from the thickener (30 to 35 percent solids by weight), the fine waste stream, is sent to a slurry impoundment. Deep cone or other paste thickeners produce an underflow with a higher solids content than a conventional thickener. Their steep-sided deep cone construction takes advantage of the high differential pressure applied by the depth of solids to produce a paste (Steve Slottee, Eimco Process Equipment Company,

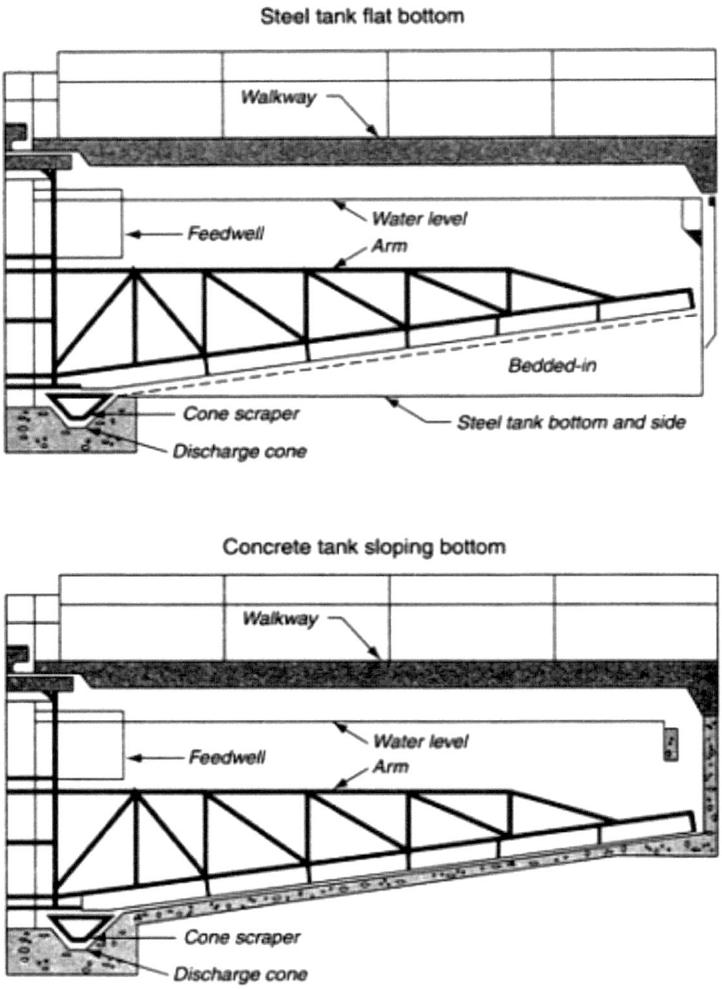

FIGURE 7.2 Thickener tank design. Reprinted with permission of the Society for Mining, Metallurgy, and Exploration, Inc., www.smenet.org.

personal communication, 2001). Deep cone thickeners produce dewatered paste in the range of 65 to 75 percent solids by weight (Osborne, 1988). While this material still requires disposal, its volume is less than that of unthickened slurry.

Four types of filtration devices are used in coal preparation: gravity, vacuum, pressure, and centrifugal. Although gravity and centrifugal methods are used extensively, vacuum and pressure filtration methods offer the greatest potential for dewatering material smaller than 100 mesh (150 micrometer) that typically leaves the conventional thickener as underflow.

Three basic types of vacuum filtration devices are rotary drum, rotary disc, and horizontal belt or disc, depending on their filter configuration type. With all of these devices, the feed is applied to the filter medium under vacuum. This draws the water through the filter medium while retaining the solids on the surface. The filter cake is then removed either by a burst of air pressure, mechanical scraping, or both. Vacuum filters produce a filter cake with a moisture content of 20 to 30 percent (Leonard, 1991).

Pressure filtration devices are classified as either batch or continuous press. Batch devices, such as the plate and frame filter press, have not been widely accepted in the U.S. coal industry because they operate discontinuously (Osborne, 1988). However, solids recovery is high and effluent water is clear.

Belt filter presses operate continuously and are more widely used than batch methods (Figure 7.3). Belt presses produce a cake with a moisture content reportedly in the range of 20 to 30 percent (Osborne, 1988); however, in practice, moisture content ranging from 35 to 40 percent is more common. Belt filter presses currently produce a dewatered product that still must be disposed of behind a retaining structure.

Hyperbaric pressure filters, a fairly recent development, combine the filtration technology of disc or drum filters and the low moisture levels of discontinuous plate and frame presses. The specific solids throughput of a hyperbaric filter is several times that of a vacuum filter or other batch-operation pressure filters (B.K.Parekh, University of Kentucky, personal communication, 2001).

Chemical additives are almost always used in conjunction with any dewatering mechanism. Most of the mechanical methods (thickeners, filters, and presses) rely on chemical additives (usually flocculants) to expedite and enhance the separation process. The additives represent a significant operating cost, but dewatering would be largely ineffective without them.

Thermal drying is an effective way of removing moisture, but it is very expensive and energy intensive and is currently used only for drying fine coal. In addition, it entails the environmental cost of air pollutants produced

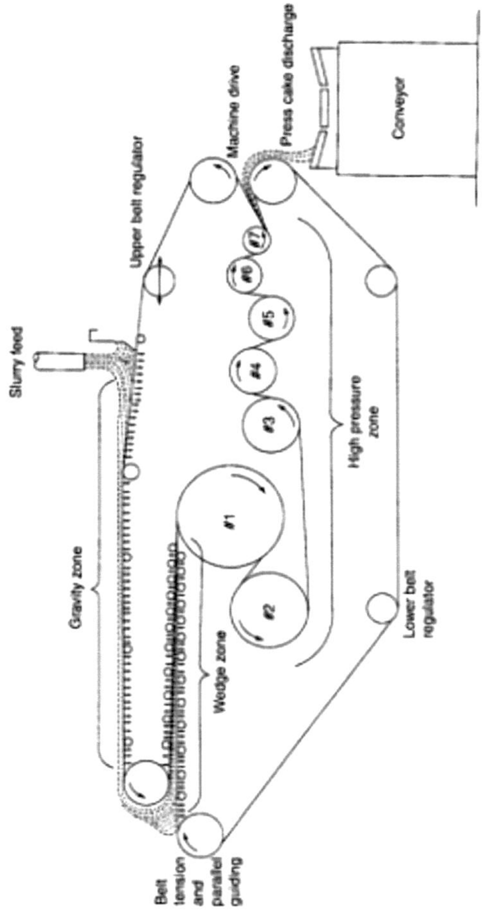

FIGURE 7.3 Belt filter press. Courtesy of Andritz AG.

during the drying processes. Other dewatering methods being developed include ceramic capillary filters, electro-acoustic dewatering, microwave dewatering, vacuum pressure hybrid filters, and pulsating vacuum filters (B. K.Parekh, University of Kentucky, personal communication, 2001).

Dewatering technologies represent one alternative to reduce the volume of waste deposited in coal slurry impoundments, but they do not eliminate the need for an impoundment. Many dewatering technologies are currently available for specific applications, though none is likely to be universally applicable. The committee believes that equipment vendors' current research and development will lead to improvements in these technologies and that operators of coal waste impoundments should monitor them carefully.

DIRECT UTILIZATION OF SLURRY

Slurry refuse can be utilized directly for power generation, either in conventional boilers or with advanced combustion and gasification technologies. These technologies have the advantage of turning a waste into a resource. Some of them can reduce cleaning requirements for coal in the preparation plant, but the use of low quality coal feed will increase the amount of waste generated at the power plant.

Conventional Pulverized Coal-Fired Boilers

Direct utilization of fine coal waste in conventional pulverized coalfired boilers is an important alternative to its disposal in an impoundment. This does not require a significant change to the system of mining, cleaning, and burning coal. However, use of this material presents significant challenges to existing boilers, because the moisture content is high and the heating value and trace element quantities are inconsistent (Harrison and Akers, 1997). If the fine waste material alone does not meet the required specifications for end-users, it can either be combined with a variety of other feeds, such as cleaned coal or biomass, to achieve the desired fuel consistency or be agglomerated to improve handling.

Fine coal is difficult to handle, even when dewatered, because it clogs equipment and is dusty and explosive. Agglomeration technologies can reconstitute the fine coal by briquetting, pelletizing, or extrusion and can solve the handling and transportation problems. The advantages and disadvantages of the various forms of agglomeration are summarized in Table 7.1. Which agglomeration technology is appropriate depends on the

TABLE 7.1 The Advantages and Disadvantages of Agglomeration Technologies

Agglomeration Method	Advantages	Disadvantages	Other Comments
Mixer agglomeration	Simplest technique; Cheapest; Simple binders (such as water) can be used	Weakest product; Converts dust to crumb-size product	Possible application to condition coals for nearby use
Disk and drum pelletizers	Simple concept; Next cheapest; Can make pellets	Relatively weak product	Can make pellets 5 to 80 mm in diameter
Roll press	Uniform product size	Relatively expensive; Needs good binders	
Extrusion	May not need a binder	Relatively expensive; Strength is problematic	Brick-shaped product is typical

SOURCE: Whitehead, 1997.

nature of the coal, required product characteristics (ease of handling and strength), coal cost and product value, and binder availability and cost (Couch, 1998).

Another alternative is to create a slurry fuel from fine coal waste. In 1973, the oil embargo prompted the development of coal-water slurry technology, which consists of a pulverized, fluidized coal feed that can be transported by pipeline and used as a fuel for utility boilers or gasifiers. The high-ash (noncombustible material) fine waste produced by coal cleaning could be treated similarly. Initially, slurries containing 60 percent solids (high-solids) were tested, but more recently, slurries containing 50 percent solids (low-solids) have been tested. Low-solids slurries require particles smaller than 100 mesh (150 micrometer).

Several utilities have demonstrated coal-water slurry utilization in conventional boilers, including Penelec (later GPU Genco), Tennessee Valley Authority, and Southern Illinois Power Cooperative (J.Morrison and B.Miller, Pennsylvania State University, personal communication, 2001). Coal-fired boilers at Seward Station in western Pennsylvania use a slurry consisting of material obtained directly from cleaning plant fines and from recovered fine waste from impoundments (J.J.Battista, Cofiring Alternatives, personal communication, 2001). Masudi and Samudrala (1996) assessed a blend of 15 percent coal waste fines and 85 percent clean coal in utility boilers. Slurry fuel produced from the blend contained 46.9 percent moisture and 19.8 percent ash. After dewatering and wet milling, the final ash content of the fuel was 10.5 percent.

Although benefits of using coal slurry include reducing nitrogen oxide (NO_x) emissions by as much as 30 percent (J.J.Battista, Cofiring Alternatives, personal communication, 2001), the costs of processing and transporting are significant. Conventional utility boilers are usually far from the coal preparation plant, and pulverized coal-fired boilers may impose stringent quality demands on the characteristics of the slurry for direct burning. The coal slurry produced at a preparation plant may not meet the requirements of the power plant. Nevertheless, there is a potential for creating a market to acquire and process slurries from diverse sources to supply custom slurry to diverse customers.

The committee concludes that technologies for utilization of fine coal waste for electricity generation in conventional coal-fired power plants are available. These technologies offer near-term opportunities for the reduction of fine coal waste disposed of in impoundments. The technologies produce coal that is more expensive than cleaned coal, as a result of capital and operating costs of additional equipment, and, in the case of coal water slurry, the additional cost of transportation. However, the avoided costs of slurry impoundments must be included in the cost comparison. A definitive cost analysis, which is necessary for any cost comparison of technologies, was not performed in this study.

Alternative Combustion and Gasification Technologies

While pulverized coal-fired combustion is the dominant technology for generating electricity from coal, other technologies have long been recognized as advantageous for operating efficiency and reduced air pollution. Fluidized-bed combustion and gasification have been commercially available for at least 10 years and show promise for recovering the heat content of fine coal waste while avoiding some of the operational problems that limit use of coal fines in conventional pulverized coal-fired boilers. Because of the potential for cleaner, more efficient fossil fuels through the use of these technologies, the U.S. Department of Energy (DOE) included these two technologies in the Vision 21 program for fossil fuel options of the future. The National Academy of Sciences reviewed this program and recommended that advances in gasification technologies be pursued aggressively in the Vision 21 program and that fluidized-bed combustion research be continued as part of DOE's main program to improve power-generating technologies (NRC, 2000b).

Fluidized-Bed Coal Combustion

Fluidized beds suspend solid fuels on upward-blowing jets of air during the combustion process (DOE, 2001). This turbulent mixing of gas and solids makes chemical reactions and heat transfer more effective. While NO_x forms at 2,500°F, fluidized-bed combustion burns fuel well below that at temperatures of 1,400 to 1,700°F. In addition, a sorbent inside the boiler can capture more than 95 percent of the sulfur pollutants. The fuel flexibility makes this technology popular—almost any combustible material, from coal to municipal waste, can be burned. Also, fluidized-bed combustion can meet SO_2 and NO_x emission standards without expensive external controls.

Approximately 12 years ago, atmospheric fluidized-bed combustion crossed the commercial threshold, and most boiler manufacturers currently offer fluidized-bed boilers as a standard package (DOE, 2001). Fluidizedbed coal combustors have been called "the commercial success story of the last decade in the power generation business." More than 6 gigawatts of electricity are produced by fluidized-bed power plants operate in the United States (Arey, 1997).

A newer technology that promises greater efficiency is the pressurized fluidized-bed combustor (Figure 7.4). The first-generation pressurized fluidized-bed combustor, which has been demonstrated by a joint DOE-American Electric plant in Ohio, the Tidd Plant, uses a bubbling-bed technology. A relatively stationary fluidized bed is established in the boiler using low air velocity to fluidize the material, and a heat exchanger (boiler tube bundle) immersed in the bed to generate steam.

Currently, investigators are developing a second-generation pressurized fluidized-bed combustor to enhance efficiency. Circulating fluidized-bed technology can improve operational characteristics by using higher air flows to entrain and move the bed material, and by recirculating nearly all the bed material with adjacent high-volume, hot cyclone separators. The relatively clean flue gas goes on to the heat exchanger. This approach theoretically simplifies feed design, extends the contact between sorbent and flue gas, reduces likelihood of heat exchanger tube erosion, and improves SO_2 capture and combustion efficiency (DOE, 2001). With all these features, second-generation pressurized fluidized-bed combustion is expected to achieve a 52

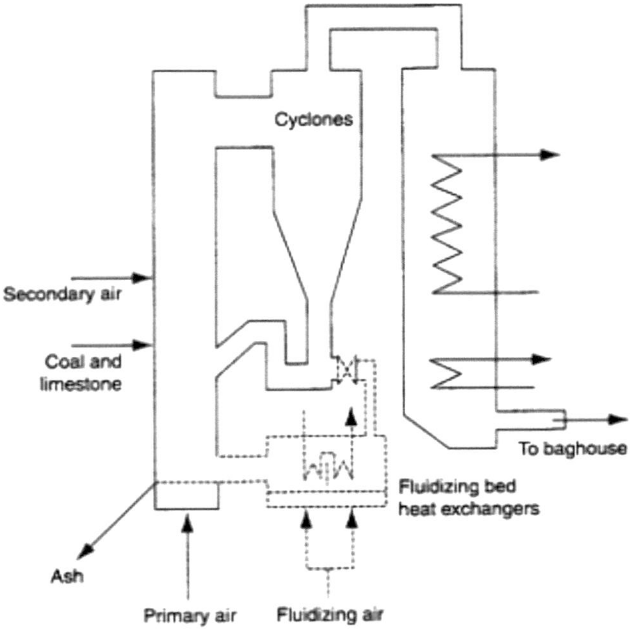

FIGURE 7.4 Typical circulating fluidized bed arrangement. Reprinted, with permission, from Rousaki and Couch, 2000. Courtesy of IEA Coal Research.

percent fuel-to-electricity efficiency level and have near-zero NO_x, SO_2, and participate emissions. Market entry is projected for 2008 (DOE, 2001). Coal-water slurry has advantages as a feed for pressurized fluidized-bed systems, as compared with dry lump coal, because it can be easily introduced by pumping (J.Morrison and B.Miller, Pennsylvania State University, personal communication, 2001).

Fluidized-bed combustion has also been demonstrated for low heat content waste found in piles in the Pennsylvania anthracite and bituminous mining regions and elsewhere. At least 14 such plants are operating in Pennsylvania (Couch, 1998), and additional plants have operated in West Virginia and Illinois. Similarly, fluidized-bed plants in Utah and Montana burn coal from mine refuse piles consisting of low-grade surface coal layers discarded during the beginning of surface mining operations (Couch, 1998;

Syngle and Sinn, 1991). Fluidized-bed coal combustion technologies are also being used in other countries (Sidebars 7.3, 7.4, and 7.5).

Fluidized-bed power plants can eliminate waste generation at the preparation plant, but they increase the waste produced during the utilization phase (Figure 1.2). In general, fluidized-bed power plants produce more coal combustion waste than conventional coal-fired boilers. However, because of the alkalinity (high pH) created by combustion of limestone, the waste can be used to reclaim acidic mining lands (Couch, 1998). Roughly 75 percent of the wastes from fluidized-bed combustion are used beneficially, primarily in mine-fill (61 percent), followed by waste stabilization (6 percent), construction fill (5 percent), and agriculture (1 percent) (EPA, 1999). In addition, coal combustion waste requires less volume for disposal and can be backfilled into mine workings (Couch, 1998).

Gasification

Gasification is the process of converting various feedstocks to fuel gas or syngas under reducing conditions, at high temperatures, with the addition of oxygen and steam (DOE, 1999; Hebden and Stroud, 1981; J.L. Johnson, 1981; Rousaki and Couch, 2000). The feedstock reacts, and the product is then cooled and purified. This technology has been in use since the 1930s (Rousaki and Couch, 2000). Feedstocks used in gasification include coal, coal slurry, petroleum, gas, petroleum coke, and biomass (DOE, 1999; Rousaki and Couch, 2000). Coal has been used as an experimental gasification feedstock since the 1970s. The recent Vision 21 review concluded that coal gasification should be a major focus for DOE's enabling technologies program (NRC, 2000b).

Incorporating gasification with other technologies has resulted in the integrated gasification combined cycle, which improves energy conversion processes by combining gasification and gas cleaning, synthesis gas conversion and turbines (Figure 7.5) (DOE, 1999). This process can theoretically reach 60 percent efficiency and has been reported at 42 percent efficiency (Arey, 1997; DOE, 1999), whereas typical coal-powered plants achieve a maximum of 34 percent efficiency (Arey, 1997).

Emerging integrated gasification combined cycle power plant technologies use many different types of coal gasification reactors based on entrained beds, fluidized beds, and fixed- or moving-bed technologies. A feature to note is that all of these coal gasification reactors use steam as a reactant. These technologies could utilize coal-water slurry as a feed with

SIDEBAR 7.3 ATMOSPHERIC FLUIDIZED-BED COMBUSTORS

Firing atmospheric fluidized-bed combustors with fine coal waste from preparation plants has been demonstrated in a number of countries—Australia, Canada, Japan, South Africa, China, and India (Couch, 1998). Waste slurry from impoundment cleanup can also be fired (Couch, 1998). This use of fine coal waste can reduce traditional coal processing steps (and costs) for mined coal, because the boilers will operate at an overall lower energy content than conventional boilers (Couch, 1998; C.Norris, Geo-Hydro, Inc., personal communication, 2001). Fluidized-bed plants are generally smaller than conventional power plants, and the boilers can be built in a factory and shipped to the site. Operating them close to the mine can result in savings in coal transportation costs. Limestone, however, as part of the fluidized bed, must be transported to the site to remove SO_2

SIDEBAR 7.4 ATMOSPHERIC CIRCULATING FLUIDIZED-BED BOILERS

Alstom Power and the Warkworth Mine in New South Wales, Australia, have demonstrated the firing of an atmospheric circulating fluidized-bed power plant with fine coal waste (J.Durant, Alstom Power, personal communication, 2001). The 140-megawatt Redbank Power Project in New South Wales uses two fluidized-bed boilers fired with 100 percent waste coal, both from an impoundment and from the mine's preparation plant. The coal waste is fed from a preparation plant using Jameson cell technology followed by centrifugation to reduce the moisture content to approximately 30 percent. The ash content of the fine coal waste is around 15 percent, and the particle size of 58 to 79 percent of the feed is smaller than 200 mesh (75 micrometer). A conveying system transports the fuel from the preparation plant to the power plant site (Goldbach and Tanca, 2001). SO_2 emissions have been estimated at 412 pounds per hour and NO_x at 397 pounds per hour of burning coal waste (J.Durant, Alstom Power, personal communication, 2001). Alstom has similar projects underway in Pennsylvania and Illinois, although these projects will use fuel of a larger particle size and lower moisture content (J.Durant, Alstom Power, personal communication, 2001).

SIDEBAR 7.5 CONVENTIONAL BOILERS AND FLUIDIZED-BED BOILERS

The largest plant In the world burning fine coal waste from a preparation plant is in France. The plant includes both a conventional boiler and a fluidized-bed boiler with a total capacity of 1,200 megawatt of electricity. The conventional boiler is fed by an 8 percent moisture vacuum-filtered material from the preparation plant, while the fluidized-bed boiler is fed by a fine coal slurry from the preparation plant and old slurry impoundments (Couch, 1998). The slurry burned by the fluidized-bed boiler contains 33 percent water, 30 percent or more ash, and a high clay content, which makes it difficult to dewater further.

The EPA reviewed the hazards of disposal of coal combustion waste, including that from fluidized-bed combustion, and concluded that these wastes should retain their exemption from hazardous waste regulations under the Resource Conservation and Recovery Act. The EPA also concluded that additional regulations for non-hazardous solid waste should be implemented under Subtitle D of Resource and Conservation and Recovery Act (EPA, 2000). The fluidized-bed coal combustion waste generally did not leach metals or toxic compounds, but the use of fluidized-bed combustion wastes for mine reclamation warrants further review because of possible groundwater consequences (EPA, 1999).

the water in the slurry generating the steam (J.Morrison and B.Miller, Pennsylvania State University, personal communication, 2001).

Low-value coal can be used but must be accommodated in the initial system design. Additional research and operational experience will be required to specify the coal property requirements for the different types of gasifiers (Rousaki and Couch, 2000). High-ash and low-moisture coals perform best in moving-bed gasifiers, in which the temperature differentials permit localized melting of ash, which can then be removed as slag. Fluidized-bed gasifiers do not handle ash well; but in entrained-flow gasifiers, ash can be converted to slag by adding a fluxing agent.

Barriers to the integrated gasification combined cycle process include high capital cost and lower reliability than operators currently require (Rousaki and Couch, 2000). These barriers reflect the complexity of process designs and limited operational experience. All U.S. integrated gasification combined cycle plants have been subsidized by the government, as they have in Spain and Germany. In addition, to produce 800 megawatts of electricity, the turbines and gasifiers must be in parallel trains, introducing a size limitation.

Despite these challenges, seven integrated gasification combined cycle plants are in use. Three in the United States (Florida, Indiana, and Nevada)

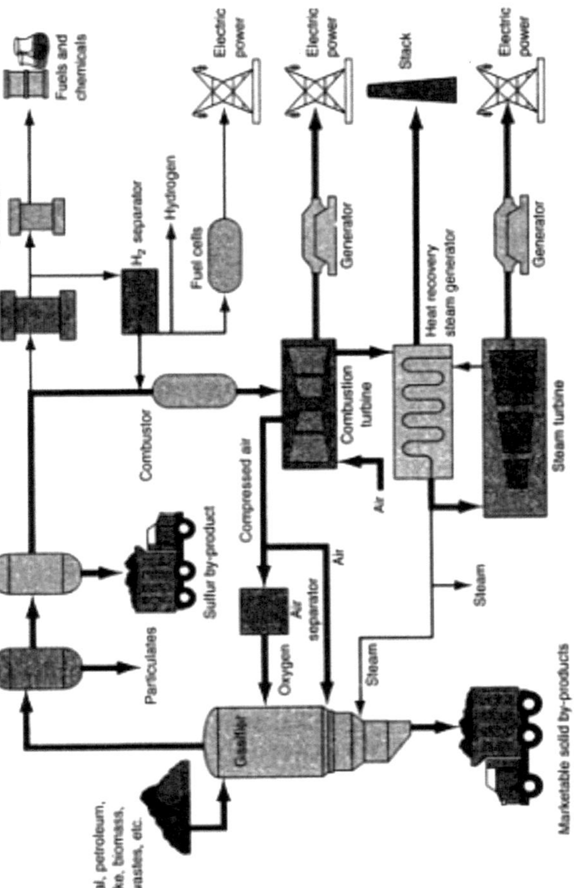

FIGURE 7.5 Integrated gasification combined cycle. From DOE, (1999).

have been in operation since the mid-1990s (Rousaki and Couch, 2000). The site in Indiana has a two-stage, entrained-flow gasifier that takes bituminous coal slurry feed. An entrained-flow gasifier in Florida uses slurry with approximately 10 percent ash. The Pinon Pine plant in Nevada uses high-quality coal from Utah in its fluidized-bed gasifier. A new plant with a moving bed gasifier is planned in Kentucky and will use high-sulfur, bituminous coal and municipal solid waste feed (Rousaki and Couch, 2000).

Gasification has significant environmental benefits, because it creates less of the traditional waste products, such as particulates and wastewater. In addition, 99 percent of the sulfur can be removed, and NO_x can be reduced by 90 percent and CO_2 by 35 percent (Arey, 1997; Rousaki and Couch, 2000). Most of the ash produced in gasification is recirculated through the system and converted to the preferable and lower volume bottom ash or slag (lyengar and Ramakrishnan, 1992). Sulfur and other trace elements can be removed from the gas stream. Waste such as elemental sulfur and vitrified slag are marketable products. In addition, less water is used in integrated gasification combined cycle plants than in traditional coal-fired plants (Arey, 1997).

Although gasification is a proven technology, costs are still prohibitive for implementation even using clean coal. However, the environmental benefits of improved emissions and potential uses for waste products appear to be significant. Research is still underway for potential markets for waste slag (Barbara Arnold, PrepTech, Inc., personal communication, 2001).

As part of the Vision 21 program, DOE recently selected two projects for the demonstration of gasification facilities capable of utilizing coal water slurry as the gasifier feed to produce a combination of electric power, heat, fuel, and chemicals. In another Vision 21 project, gasification of a biomass and coal slurries blend is being demonstrated (J.Morrison and B.Miller, Pennsylvania State University, personal communication, 2001).

The committee concludes that the combustion of fine coal waste in advanced combustion technologies, such as fluidized-bed combustion and gasification, is an alternative that also shows considerable long-term promise. Atmospheric fluidized-bed units are already in use for combustion of fine coal waste slurries from both preparation plants and old slurry impoundments, but they have not gained wide usage. The available fluidized-bed combustion units generally produce more expensive electricity than conventional boilers, but they have advantages in reducing air pollutants (SO_2, NO_x, CO_2), and their relatively small size for location near coal mines makes them worthy of the utility industry's consideration. Pressurized fluidized-bed technologies offer gains in efficiency over atmospheric technologies but have not been utilized in full-scale applications for burning fine coal waste. Gasification technologies are also promising for coal water slurries, because they operate more efficiently, and emerging technologies can utilize the water from the slurry as a steam source needed for the gasifier. DOE is already promoting advanced combustion technologies through research and demonstration projects. Further research is needed on the use of fine coal waste slurries as feeds, and incentives may be needed if these technologies are to be utilized widely for fine coal waste combustion. While coal combustion wastes from power plants are already being used for a number of purposes, the issue of the management of coal combustion waste from these advanced combustion technologies should be studied further.

ALTERNATIVES TO DISPOSAL IN IMPOUNDMENTS

While impoundments are widely used in the Eastern U.S. coalfields for the disposal of slurry from the preparation plant, there are several alternative methods of slurry disposal. Some of these methods are already in practice wherever applicable. These methods are discussed here as surface and underground methods.

Surface Methods

Throughout the United States and in other countries a number of alternative surface disposal methods have been used that do not rely on a typical cross-valley impoundment structure. Some methods are designed for slurry only, others for dewatered fine refuse, and still others for a combination of fine and coarse refuse (Table 7.2). These methods are influenced by topography, geology, and mining and coal preparation characteristics and are therefore not universally applicable.

Incised ponds function as normal slurry impoundments and are considered to be a form of impoundment. Unlike other impoundments, incised ponds do not rely on a structure (i.e., an embankment) to contain the slurry. Incised ponds can accept any form of coal waste but are commonly used to contain slurry. This method of surface disposal is most common in regions where surface mining is practiced. In the United States, their use is limited primarily to the Interior Coal Basin (specifically, Illinois, Indiana, Ohio, Pennsylvania, and western Kentucky). Surface mining produces long end-cuts and inclines, which are usually allowed to fill with water to form a permanent lake. These lakes are at or below the level of surface drainage and

TABLE 7.2 Disposal Methods

Alternative Disposal Methods	Potential Application	Advantages	Disadvantages	Other Comments
Surface Methods				
Incised pond	Interior coal basin		Dewaters slowly; Heavy metals in groundwater	
Dewatered fine refuse cells	Appalachia	Above drainage	Compaction problems; Resaturation of fines	
Combined refuse			Needs mechanical transport; Mechanical dewatering not adequate; Difficult for trucks and conveyors; Difficult to compact	
Co-disposal	Sparsely populated areas with low rainfall	Stabilizes quickly; Less volume than fines and coarse separately; Dewaters quickly	Needs starter impoundment structure	Ratio of coarse to fine critical; Proper gradation and shape
Underground Methods				
Uncontrolled placement	Where access is unavailable (abandoned mines)		Treatment of discharge water	Potential for plugging of injection system
Controlled placement	Where access to underground works is available	Can verify mine maps and accurately estimate the space; Can construct bulkheads; More practical collection and reuse of water; Protects against subsidence; Better use of space; Greater certainty in location of waste	Worker safety issues	

do not present a danger of sudden failure. Although underground mines may be below these ponds, they are generally separated by a considerable thickness, and the underground workings do not crop out above the level of surface drainage.

Where incised ponds are used, slurry is pumped to the pond, and surface and groundwater are prevented from filling the cut. The long cuts can be divided into cells with plugs of spoil material. The cells are filled sequentially, and subsequent cells can serve as overflow containment for the cells closest to the discharge point. Clear water is decanted from the last cell for recycling through the preparation plant. The pond is operated similarly to other slurry impoundments. When the pond is full, it can either become a wetland or be reclaimed by covering it with soil and revegetating the soil. Incised ponds do not have the benefit of underdrains; consequently, they tend to dewater slowly and may present problems in traversing the fill with earth-moving equipment if the reclamation plan calls for covering the surface with soil. Permits require groundwater monitoring and compliance with all surface water effluent limitations.

Slurry cells are used to dispose of waste in a diked impoundment. This type of impoundment reduces some of the problems and risks associated with the more common cross-valley impounding structure. Berms that bound individual cells are constructed of compacted coarse refuse. Cells may reach depths of 12 feet, with 8 feet of fine refuse covered by 4 feet of compacted coarse refuse; if they are designed to impound less than 20 acre-feet of slurry, they do not have to be permitted as an impoundment (M.Day, Arch Coal Company, personal communication, 2001).

Slurry from the thickener is pumped directly to each cell. After a cell fills, the discharge point is moved to another cell. Water can be decanted from the surface, or the cells can be allowed to dewater by evaporation. The time required varies according to weather conditions, but the fine refuse is usually dry enough in two or three months to allow coarse refuse to be placed on top and compacted. Subsequent cells can be constructed on top of completed cells to a final height of a few hundred feet. For stability, benches must be constructed every 50 feet to reduce the overall slope angle.

The advantage of this method over conventional cross-valley slurry impoundments is that each cell is small and self-contained and can be designed according to the strength properties of the coarse refuse. The main disadvantage in steep terrain is the limited availability of flat land to construct the cells. Another disadvantage is that slurry cell operations are not compatible with a high production rate at the coal preparation plant. The maximum plant capacity for this type of disposal option is about 500 tons per hour (E.Kitts, Summit Engineering, personal communication, 2001).

Even at this rate several cells must be permitted and in varying stages of operation at any one time.

A variation of slurry cells is dewatered fine refuse cells. Mechanically dewatered fine refuse can be placed in bermed cells. However, the combined costs of dewatering and cell construction prevent this method from frequent use in the coal industry.

Combined refuse disposal is another option. Combined refuse refers to fine refuse from the static thickener that has been mechanically dewatered and combined with coarse refuse for disposal on the surface. Depending on the percentage of coarse and fine refuse and the moisture content of each, the moisture content of combined refuse will range from 15 to 20 percent. This material must be transported mechanically, since it is too dry to be pumped. It is deposited in lifts approximately 2 feet thick, then graded with a bulldozer, and compacted to create a stable surface. Once the fill reaches the designed level, it is covered with 4 feet of soil and revegetated. Combined refuse fills require underdrains to capture any water that leaches from the fill, and perimeter drains to prevent runoff from entering it.

Problems associated with this method of disposal include the frequent inability of mechanical dewatering to produce the moisture level required for disposal of a combined refuse material (David Carris, J.T.Boyd Company, personal communication, 2001). Because the moisture and clay content are still fairly high, trucks or conveyors cannot handle the material easily, and compaction is difficult. Davies and others (1998) report using small amounts of cement to create a more stable refuse material. They found that the addition of 2 to 4 percent cement, by weight, results in a material that can be compacted using standard procedures. Small amounts of fly ash and lime have also been used to stabilize combined refuse. In addition, this method is fairly expensive because of the high cost of mechanical dewatering and the potential need for stabilizing chemical additives. The need for additives may be seasonal and depends on climatic conditions. Finally, the method is best suited to flat land.

Co-disposal, developed and practiced primarily in Australia, involves the combination of fine refuse from the static thickener with coarse refuse (Williams et al., 1995). Since the fine refuse is not mechanically dewatered, the combination has a fairly high moisture content that allows the combined material to be pumped to the disposal area. The main difference between this material and conventional slurry is the presence of coarse refuse in the mix. The advantages are that it requires less total storage volume than separate fine and coarse disposal methods, and the refuse stabilizes more quickly than typical slurry.

However, co-disposal does not eliminate the need for a starter impoundment structure. After an initial dam is built, the mixed beach sediments, settling out of the co-disposal material can be used to construct the embankment for above-ground storage. Ideally, the fine particles fill the voids between the coarse refuse fragments. As the mixture is discharged, it forms a steep beach sloped at approximately 10 percent. This slope decreases to approximately 1 percent at the end of the mixed beach. At this point, the sediment consists mainly of fine coal suspended in ponded water in a small impoundment. The success of this method is closely linked to the ratio of coarse to fine particles. It also depends on gradation of the refuse (large gaps in particle size are not acceptable) and proper particle shape (angular or "platy" particles cause problems). Since the impounding structure is raised by deposition of mostly coarse material, it does not compact as the structure increases in elevation.

This method has been used primarily in sparsely populated areas with low annual rainfall. Questions remain about its suitability for steep hills with high annual rainfall. Unlike a conventional slurry impoundment, which contains only fine refuse, the co-disposal system places all refuse (both coarse and fine) in a slurry and deposits it behind an impounding structure. Therefore, even though the refuse dewaters more quickly and forms a stable bench, it requires more impoundment storage volume than an impoundment designed only for fine refuse. So, for steep terrains, this factor negates the advantage of less total storage area by actually requiring more material (both coarse and fine) to be placed in an impoundment. Its use would hinge on whether increased stability of the refuse outweighs the additional volume of the impoundment.

If an effective dewatering approach, such as paste thickening, is used, the resulting waste can be disposed of by thickened high-density residue stacking (Lech Brzezinski, LSB Consulting Services, personal communication, 2001). Deep cone paste thickeners produce a homogeneous, non-segregating paste with a solids content of approximately 60 percent. The degree of dewatering is determined by the pumping capabilities. Under controlled conditions, the paste can be deposited in thin layers over the disposal site at uniform slopes of 2 to 5 percent and does not require an impoundment structure. This method is most suitable for homogeneous residues of fine gradation, where the thickening process prevents segregation of the coarse and fine particles during transportation.

Thickened high-density residue stacking was developed more than 20 years ago to handle red mud tailings generated by alumina plants. It has been used for approximately 10 years for disposal of gold and base metal tailings;

has recently found application at power plants for disposal of ash; and has seen limited use in the coal industry for the disposal of fine coal refuse.

Although rainfall does not reconstitute a properly formed paste, and erosion is not excessive on typical stacks with 2 to 5 percent slopes, this method is best suited to areas of low rainfall and high evaporation. Under these conditions, the surface of the stack can be traversed by bulldozers within days of final placement. Regardless of the amount of precipitation, perimeter drains are necessary to catch runoff and divert it around the stacks. Underdrains are also needed for stability to prevent an increase in the phreatic surface.

Three considerations—land availability, steep terrain, and cost—hamper applying unsupported thickened high-density residue stacking to fine coal refuse disposal. This method is best suited to areas where the slope of the land is less than 5 percent. Secondly, although the technology is based on a significant modification of the standard thickener, these thickeners are much deeper and require a longer residence time. Therefore, the lower throughput rate of deep cone thickeners compared with that of standard thickeners may significantly affect the economic feasibility of the method.

Underground Disposal

In 1975, the National Research Council conducted a study entitled *Underground Disposal of Coal Mine Wastes,* which evaluated the technical and economic feasibility of several underground disposal methods; namely, pneumatic backfilling, hydraulic backfilling, mechanical backfilling, hand packing, controlled flushing, blind flushing, and pneumatic flushing. Much of the study drew heavily on European experience, where backfilling has long been used to control subsidence. The backfilling methods emphasized the use of coarse refuse and did not specifically address the issue of fine refuse or slurry; only the hydraulic methods dealt specifically with fine refuse. The report did not discuss pneumatic flushing of fine coal waste, but it did indicate that this method had been used to inject fly ash into underground mines. The two primary methods for injecting fine coal refuse into underground mines are controlled flushing, where the underground workings are accessible, and blind or uncontrolled flushing, where the underground workings are abandoned or have caved in.

The 1975 NRC report found that a number of underground disposal methods were technologically feasible at that time; however, none was universally economically feasible. The optimal solution varies from site to site. In addition, the question of workers' health and safety was raised when considering injection into active mines. The report recognized that safe and nonpolluting disposal is the operator's responsibility. If regulations are imposed that exceed this obligation, society must recognize its responsibility to avoid inequitable distribution of added costs to some operators. Under the right set of conditions, underground disposal of slurry is an attractive alternative to impoundments and is currently being used where suitable opportunities exist (Table 7.2).

Slurry injection creates additional factors that must be considered. Many issues related to underground injection of slurry are independent of the method of slurry injection. For example, it is essential to have an adequate supply of water. This is especially true when water is not being recaptured from the underground workings. Also, it is important to keep the solids content below 20 percent, and preferably in the range of 10 to 12 percent. Additional common issues include surface ownership, permits, surface layout, and surface drainage.

In most cases, the right to mine coal is obtained after leasing the mineral rights from the owner, but often, the rights to the surface have been severed. Therefore, a lease to mine the coal does not automatically entitle the mine operator to inject coal waste back into the voids created by the mining process. Unless the mining company owns both the surface and the mineral rights, the operator must obtain the landowner's permission to inject coal waste back into the mine. In Kentucky, the state regulatory authority has determined that a company cannot inject coal refuse into an underground mine for which it has the mineral rights unless all of the surface owners above the underground mine approve of the plan. In areas where multiple ownerships are common, this policy seriously limits the areas available for underground injection.

Under SMCRA, each state's regulatory authority has responsibility for issuing permits to inject coal waste into underground mines. The permit application must address such issues as:

- Source and quantity of waste,
- Area to be backfilled and the method to be used,
- Approximate percentage of the voids that will be filled,
- Design of underground bulkheads,
- Potential influence on any active underground mines,
- Source of the water to be used,
- Methods of dewatering the backfill,
- Amount of water that will remain underground, and
- Water treatment system that will be used.

Underground injection also requires MSHA's approval. Depending on the state, this can be handled independently or through a joint approval process. Finally, underground injection of slurry requires approval by the regional EPA office, whose primary concern is groundwater quality. States may interpret the requirements of the Safe Drinking Water Act differently— for example, West Virginia limits the injection of any slurry from circuits that employ petroleum products, such as those used in froth flotation.

In addition to obtaining permission from the surface owners to install injection wells, the operator must obtain the right-of-way for the pipeline to deliver the slurry to the boreholes (Marshall Hunt, Consol, personal communication, 2001). Variable terrain can impose severe pumping problems if pipelines must be laid over hills and down into valleys to access injection sites. The accuracy of mine maps, with regard to underground workings and surface surveys, is important if the slurry boreholes are to intercept the underground workings in the desired location.

Filling above-drainage mine workings with slurry may increase hydraulic head on the coal barriers and result in a blow-out, making evaluation of mine workings above a surrounding stream valley critical. The mine maps must be evaluated for accuracy, and the underground barriers for adequacy to contain the slurry. Mines below the surrounding natural drainage level offer more secure underground disposal sites.

Blind flushing is used in mines where access has been obstructed, such as by roof falls. In this method of slurry injection, the underground cavity may be dry, partially filled with water, or completely filled with water. The volume of the voids is estimated from old mine maps and any other available, relevant data. Since it is nearly impossible to determine how much slurry a borehole will accept, a series of holes is drilled; when one borehole becomes clogged, injection moves to the next. The slurry is pumped to the borehole and injected into the mine at a relatively high velocity. Once the slurry leaves the turbulent area at the bottom of the borehole, the coarser material will settle.

In flooded mines, the injected slurry will displace the water, which often results in a new or increased discharge elsewhere (Marshall Hunt, Consol, personal communication, 2001). Depending on the quality of the displaced water, treatment facilities may have to be upgraded to handle the additional volume, or the mine pool may have to be pumped to avoid discharge at an undesired location.

Paste backfilling has been demonstrated in other types of mines. Although it has not been used to dispose of coal waste, it may be possible to extend the technology to this application.

Controlled placement is used where underground workings are accessible. This method has a number of inherent advantages over blind flushing. First, the engineer can verify the accuracy of the maps, inspect the condition of the openings, and better estimate the available space. Second, and possibly more important, bulkheads can be constructed to control the direction and extent of flow. Finally, collection and reuse of the water is more practicable.

Once areas have been identified and properly confined, controlled flushing can proceed similarly to blind flushing. However, in controlled flushing, the slurry can be delivered to the underground opening by a pipeline that enters the mine through a borehole or other opening and then extends laterally through the mine to the desired location. This allows the operator to begin filling the most remote part of the mine and then to withdraw the pipeline in stages as the voids are filled and sealed. The advantages of this method are greater certainty of the location of the slurry, better utilization of available space, and better reuse of the water needed to transport the refuse. This method also protects against subsidence. One of the main concerns is the safety of workers in the mine receiving the slurry and in any adjacent, down-dip mines.

The committee concludes that although there are alternatives to disposing of coal waste in impoundments, no specific alternative can be recommended in all cases. Acceptable alternatives are highly dependent upon regional and site-specific conditions. Also, the alternatives that have been identified are in varying stages of technological development and implementation. One of the factors limiting implementation to this point has been the cost associated with the various alternatives. Additional research is needed to develop these alternatives further and to evaluate the economics of these processes. **The committee recommends that a screening study be conducted that (1) establishes ranges of costs applicable to alternative disposal options, (2) identifies best candidates for demonstration of alternative technologies for coal waste impoundments, and (3) identifies specific technologies for which research is warranted.** Input from MSHA and OSM regarding regulatory issues will be valuable to such a study.

REMINING SLURRY IMPOUNDMENTS

While coal waste impoundments have generally been viewed as permanent disposal sites, there may be situations in which impoundments can be a resource (Sidebar 7.6). In the past, recovery of fine coal was not as efficient, resulting in many older slurry impoundments containing significant

SIDEBAR 7.6 POND RECOVERY FACILITY

Ginger Hill is a coal slurry recovery facility in western Pennsylvania that produces approximately 300,000 tons per year of synthetic fuel (Akers et al., 2001). A dredge is used to extract the fines from the impoundment. The slurry is pumped to a surge tank and then into the cleaning plant. The coal is classified at 150 mesh (106 micrometer) using 15-inch classifying cyclones. Particles larger than 150 mesh (106 micrometer) are cleaned using water-only cyclones and spirals, with an additional classification of the fine coal product by a two-stage Vari sieve. Material smaller than 150 mesh (106 micrometer) material is classified using a 6-inch classifying cyclone at 270 mesh (53 micrometers). The 150 x 270 mesh (106 x 53 micrometer) fraction is cleaned by column flotation. After cleaning, all clean coal size fractions are mixed and dewatered by two 40-inch decanter screen-bowl centrifuges. The clean fines are pelletized using COVOL binder.

amounts of coal refuse larger than 28 mesh (600 micrometer) with recoverable energy value. As processing technologies and the capacity of dewatering equipment have improved, the proportion of particles smaller than 100 mesh (150 micrometer) in slurry refuse in impoundments has increased. In many instances, the finer slurry materials being disposed of today with less recoverable and marketable coal can close off access to the more amenable, profitable, and recoverable slurry (Sidebar 7.1).

If an impoundment contains at least 1 million tons of in-situ slurry, a recovery rate of at least 30 percent of a marketable fine coal product (300,000 tons) from the slurry could prove to be a profitable venture. Fine coal from is recovered from an impoundment through several stages: investigation (preliminary site investigation, sampling and analysis of the slurry and embankment materials, and engineering design and economic evaluation), excavation and transport, and fine coal recovery.

The preliminary site investigation involves inspecting the impoundment and reviewing maps and any other available information. The sampling and analysis of slurry and embankment materials are the most critical and most expensive phase of investigation. The preliminary sampling program requires approximately three samples from each embankment and a detailed core sampling plan throughout the basin. Samples are analyzed for size distribution, float and sink characteristics, froth flotation, and percentage of moisture, ash, sulfur, and energy value per pound. In addition to these standard analyses, it may be advantageous to determine whether other materials, such as magnetite, may be recovered from the slurry. During engineering design and economic evaluation—the final phase of investigation—the economic feasibility is assessed for the project, and the marketability of the fine coal product (i.e., potential buyers, final destination for the fine coal product, means of transport) is investigated.

The excavation of the slurry material from the impoundment and delivery to the fine coal recovery plant is the most critical consideration in the development of a fine coal recovery project. Regardless of the simplicity or sophistication in the design of the fine coal recovery plant, the slurry delivered to the system requires a consistent feed rate (tons per hour) and percentage of solids concentration (weight of slurry and water) to maintain efficiency.

Three methods for coal slurry removal and delivery systems are used: conventional excavation methods; stationary pump and wash-down systems; and dredge operations. Conventional excavation and earth-moving equipment includes earth-mover pans, excavators, backhoes, drag-lines, front-end loaders, dump trucks, and farm tractors equipped with discs or harrows. The stationary pump and wash-down system agitates, washes, and keeps in suspension the slurry material within the area around and directed to the suction of the floating agitator-feed pump. Dredging uses a floating pump with a movable suction device. It excavates along a side-to-side arcing or slewing motion with a positive forward movement, cutting materials from beneath the surface of the water and creating a slurry to be delivered to the pump suction. During removal of the fine coal, the site is vulnerable to storm water runoff and to slumping of surrounding slopes, and adequate diversion structures must be used to route excess water from the facility.

The design of a fine coal recovery plant is the same as that in conventional coal preparation plants and typically processes the slurry material that is smaller than 3/8 inch. The recovery plant includes sumps, pumps, vibrating screens, sieve bends, cyclones, spirals, flotation cells or columns, centrifuges, conveyors, and thickeners—each unit system being of sufficient size and capacity to handle a specific feed rate and the size characteristics of the slurry material. The refuse from a remining operation can be handled with a variety of methods that are site specific because each impoundment has its own design, site conditions, and slurry materials. Methods include those outlined in this chapter such as underground injection, thickening of slurry, multiple cell construction deposition within the same impounding structure, a combination of these listed, or simply disposing of the refuse in a separate part of the impoundment being remined.

Recovery of fine coal from slurry impoundments is an established practice. *The committee concludes that as advances are made in the use of low value coal or coal water slurry, remining of slurry impoundments can be an attractive source for fuel supply.*

IMPLEMENTATION

Coal waste impoundments are part of the system for mining, preparing, and combusting coal for energy production (Figure 1.2). In order to assess alternatives thoroughly, the whole system of mining, preparation, refuse disposal, transportation, and power generation should be explored through an in-depth life-cycle assessment, including cost assessment, with the goal of optimizing the system to generate less fine coal waste while maintaining the performance and economics of the system. Only through such modeling can the benefits of system integration be fully explored. Of course, even if there are such benefits, they may be difficult to realize because of differences in interests and perceptions between the mining industry and the utility industry, and the resistance to change embedded in these mature industries.

Several types of policies can effectively promote the use of alternative technologies. One approach is to phase out the dominant technology, providing sufficient time for commercialization of alternatives. Such an approach can have a significant economic impact on the industry affected. A "safety valve" can be provided by extensions to the phase-out dates where alternative technologies are not available and by site-specific variances. This approach was used effectively in the Hazardous and Solid Waste Amendments of 1986 (42 U.S.C. § 6924) to phase out the landfilling of untreated hazardous waste and to give a clear signal to the market for alternative treatment technologies. This approach has the benefit of stimulating the market for alternatives without necessarily promoting specific technologies. To bring on the alternatives, this approach can be combined with other policies, such as financial incentives and research, development, and demonstration.

A second type of policy for promoting alternative technologies is the use of financial incentives. These can take the form of tax credits or direct government subsidies. Such subsidies can be paid for by a fee on the activity being discouraged, such as a few cents per ton of fine coal waste disposed of in slurry impoundments, or through a general fee on coal mining, like the Abandoned Mine Land program under SMRCA. Both the alternatives for refuse disposal and increased use of finer coal products will aid in the reduction of slurry impoundment quantity and additional and future areas required for refuse disposal. These incentives may also be extended to companies that reprocess, reduce, or relocate the slurry materials within an existing impounding structure through the use of alternative technologies, techniques, or methods. If this type of operation is utilized in abandoned or "grandfathered" slurry impoundments, an additional incentive could be granting reduced bonding or regulatory obligations and moneys from Abandoned Mine Land funds for environmental cleanup.

A third type of policy is research, development, and demonstration funded wholly by government or jointly with the industry. The federal government, and particularly the DOE, has long had research and development programs in this area, and many of the alternatives discussed above were first proven in these programs. There is always a gap between research and development and widespread implementation of technologies, but greater emphasis on joint large-scale demonstrations with industry can help bridge this gap.

A fourth approach, particularly useful where alternative technologies are already commercially available, is to assist the industry in evaluating these alternatives in an integrated, objective, and thorough manner. Typically, information about alternatives comes to the industry piecemeal through vendors of new technologies. Rarely is it possible to perform an objective evaluation of the full range of alternatives on the same basis. If such an evaluation is performed by one company, it is rarely shared with others in the industry. The EPA's Design for the Environment Program has developed an approach for evaluating and comparing the relative costs, performance, and health and environmental risks of alternative technologies that provide the same functions. In this program multiple companies and vendors participate and supply information to an independent academic institution that evaluates the technologies. Costs are assessed using activity-based cost accounting methods that allocate the full costs of technologies, bringing costs out of overhead accounts that may have been overlooked in a traditional cost estimate. Other stakeholders, such as government agencies, labor organizations, and environmental groups, are involved in setting up the evaluation and advising the team throughout. The academic institution can keep proprietary data confidential and out of the hands of regulatory agencies. The results of the evaluation are then widely disseminated to members of the industry for their use in choosing technologies. EPA neither makes these decisions nor recommends technologies. **The committee recommends that the total system of mining, preparation, transportation, and utilization of coal and the associated environmental and economic issues be studied in a comprehensive manner to identify the appropriate technologies for each component that will eliminate or reduce the need for slurry impoundments while optimizing the performance objectives of the system.** *The committee concludes that a similar analysis of the waste use and disposal technologies that make up the coal system would have value.* **The committee recommends incorporating life-cycle assessment of the costs and environmental impacts of the alternatives to evaluate**

them on a more objective, comprehensive basis. In addition, a detailed analysis of the economic and environmental impact of the various policy alternatives should be performed.

A combination of policies can encourage the use of alternatives to coal slurry impoundments. If there is no consensus on the need to phase out slurry impoundments, financial incentives for alternatives and research and development programs can expedite the switch to alternatives. In any event, a thorough, systematic understanding of the whole system through life-cycle assessment is needed for evaluating alternatives, as is the comparison of alternatives. One of the factors limiting implementation to this point has been the cost associated with the various alternatives. Additional research and development is needed to refine these alternatives and to demonstrate their economic implementation. **The committee recommends that the use of economic incentives be explored as a way of encouraging the development and implementation of alternatives to slurry impoundments.** The development of incentives should be based on the full range of the portfolio of technologies as well as the economics of the technologies. The incentives should be linked directly to the reduction in slurry production of the utilization of slurry.

8

Conclusions and Recommendations

The charge to the committee includes three major components. First, the committee was asked to examine engineering practices and standards currently being applied to coal waste impoundments and to consider options for evaluating, confirming, improving, and monitoring the various barriers that retain coal waste material within impoundments. Second, the committee was charged with evaluating the accuracy of mine maps and exploring ways to improve surveying and mapping for underground coal mines with the goal of delineating more accurately how underground mines relate to current or planned slurry impoundments. The third committee task was to evaluate alternative technologies that could reduce the amount of coal waste generated or allow productive use of the waste. This chapter summarizes the committee's conclusions, outlines the rationale for those conclusions, and reviews the recommendations that follow from them.

ENGINEERING STANDARDS, BARRIER STABILITY, AND MONITORING

The regulatory structure that governs the design process is outlined in Chapter 2, and a detailed review of the current engineering practices used in the design of coal waste impoundments is given in Chapter 3. A substantial regulatory structure exists, and a detailed regulatory review process that covers many key aspects of the design, construction, and operation of coal waste impoundments is in place. The review of the impoundment basin has been less detailed and rigorous, and the series of incidents that involved releases of coal slurry material from impoundment basins (see Sidebars 1.3 to 1.11) indicates that more investigation of the potential for loss of integrity of an impoundment in the basin area is appropriate. Special attention should be given to the potential for breakthrough of coal slurry into underground coal mines. The authority for review of basin characterization and design appears to be covered in general language authorizing investigation of all relevant issues with respect to the impoundment. **The committee recommends that MSHA and OSM should have clear authority to review basin design.** It is not evident to the committee whether specific legislation to authorize more detailed examination of basin issues is required or whether these issues can be handled by additional rulemaking under existing authority.

The regulatory approach taken by MSHA and OSM with respect to the design, construction, and operation of the embankment portion of impoundments has been effective in general. The incidents that have occurred involving impoundments have generally not been the result of significant failures of the embankment structure. Similar guidance should now be provided for the basin portion of impoundments. **The committee recommends that MSHA and OSM develop and promulgate guidelines for the site evaluation, design, construction, and operation of basins.** They should be comparable in scope to the guidelines used in embankment design. The guidelines should include methods for mitigation of any potential pathways for release of coal slurry. Many of these mitigation measures will involve established procedures for grouting, sealing of fractures, or construction of embankments that line the basin rim, for example. At the same time, the question of engineered bulkhead barriers used to isolate underground coal mines from the coal slurry in the basin should be addressed. Current regulations specify the design of bulkheads that are intended to protect miners from underground explosions, but they do not provide for bulkheads intended to support the high hydraulic heads that can arise in coal waste impoundments. **Hence, the committee recommends that MSHA review its current practice and develop guidelines for the design of bulkheads intended to withstand hydraulic heads associated with slurry impoundments.**

While the embankments designed and constructed under the current regulatory system have generally performed according to design, the committee believes that prudence requires that MSHA and OSM continue to evaluate worldwide experience with impounding structures, and to stay abreast of lessons learned from failures experienced in other mining applications so that their design criteria reflect the latest experience in all the mine sectors. **The committee recommends that MSHA and OSM continue to adopt and promote the best available technology and practices with regard to the site evaluation, design, construction, and operation of impoundments.** MSHA and OSM should commission periodic reviews of existing technical procedures and practices, with particular attention to the basin. Results of the reviews should be disseminated to industry. Based on the outcome, MSHA and OSM may need to revise guidelines that establish minimum expectations and levels of investigation for site characterization, design, construction, operation, and closure of coal refuse impoundments.

If underground mine workings are near the basin area of an existing or proposed impoundment, it is important to assess the status of coal barriers that separate the mined-out area from the surface. Currently, no federal regulations address the amount of outcrop coal barrier to be left or the depth and nature of the overburden that is required. OSM has studied the problem of outcrop barriers but has not published conclusions to date. **The committee recommends that MSHA and OSM jointly pursue the issue of outcrop coal barrier width and overburden thickness and its competence and develop minimum standards for them.**

Procedures in place for monitoring embankment structures, which include visual inspections and instrumentation, appear to be performing as envisioned in the regulations MSHA implemented. For monitoring to be successful, it should be applied to all potential failure modes. The committee believes, however, that there are opportunities for additional, continuous monitoring that may offer timely warning of potential failure of an embankment or basin. **The committee recommends that MSHA and OSM consider requiring additional continuous monitoring in specific instances and evaluate automation of monitoring instruments.**

SITE CHARACTERIZATION

The committee examined two parts of the question of site characterization, the accuracy of mine maps and the use of geophysical techniques to delineate the extent of underground mine workings in situations where maps do not exist or may not be sufficiently accurate.

The committee notes that the accuracy of mine maps has improved with the use of modern surveying equipment and practices, but also concludes that additional improvements in surveying practices, recording information on maps, and storage of maps and related information are necessary and possible. **Therefore, the committee recommends that MSHA work with OSM and state agencies to establish standards for mine surveying and mapping. These should include the following:**

- **Determining surface coal outcrop locations by aerial topographic measurements, where adjacent to existing or proposed refuse impoundments,**

- **Implementing a coordinated and assertive approach to collecting and archiving mine maps,**
- **Scanning paper copies of mine maps into electronic data files upon receipt,**
- **Setting standards for minimum closure error for all underground closed-loop surveys and that a closed-loop survey be maintained within a standard distance (to be determined by MSHA),**
- **Recording the depth of the last cut taken to a level of accuracy to be determined by MSHA,**
- **Using state plane coordinates or latitude and longitude and bottom-of-seam elevations as the map base reference,**
- **Listing of appropriate coordinate transformation equation(s) on the mine map,**
- **Adding a qualifying statement to accompany any coordinate transformation that is based upon the alignment of surface features,**
- **Improving and maintaining the location of surface controls,**
- **Determining which mine permit documents should be retained, in what form, and for how long,**
- **Avoiding the use of coal seam names as the sole basis for determining the vertical location of an abandoned mine.**

In situations where no mine maps are available or there is reason to doubt the accuracy of maps that do exist, additional investigation of the relative location of underground mine workings with respect to an existing or proposed impoundment is warranted. Chapter 5 reviews a variety of geophysical techniques that can be used to obtain additional information. Some drilling of boreholes is likely to be of value in most site characterization efforts, but the use of geophysical techniques along with drilling has potential to provide additional useful information. The committee concludes that geophysical techniques have been underutilized in the coal-mining industry and could benefit from additional research. **The committee recommends that demonstration projects using modern geophysical techniques be funded, and that the results be widely conveyed to the mining industry and to government regulatory personnel through workshops and continuing education.** Continuing education could include the opportunity to attend short courses and seminars that present the latest technology along with case histories to support its use.

ALTERNATIVE TECHNOLOGIES

Coal waste impoundments are part of the system for mining and processing coal to produce energy. In order to assess alternatives, the whole system of mining, preparation, refuse disposal, transportation, and power generation should be explored through an in-depth life-cycle assessment, including cost assessment, with the goal of optimizing the system to generate less fine coal waste while maintaining the performance and economics of the system. **The committee recommends that the total system of mining, preparation, transportation, and utilization of coal and the associated environmental and economic issues be studied in a comprehensive manner to identify the appropriate technologies for each component that will eliminate or reduce the need for slurry impoundments while optimizing the performance objectives of the system.** The committee concludes that a similar analysis of the waste use and disposal technologies that make up the coal system would have value. **The committee recommends incorporating life-cycle assessment of the costs and environmental impacts of the alternatives to evaluate them on a more objective, comprehensive basis. In addition, a detailed analysis of the economic and environmental impact of the various policy alternatives should be performed.**

The opportunities for reducing slurry volume include mining alternatives and coal processing alternatives. However, modern methods of surface and underground coal mining offer only a limited possibility for quality control during mining. Slurry volume can be reduced by improving fine coal recovery, minimizing the mass of solids for disposal, and dewatering. Many dewatering technologies are currently available for specific applications, though none is likely to be universally applicable. The committee believes that the research and development currently being performed by equipment vendors will lead to improvements in these technologies.

Slurry refuse can be utilized directly for power generation, either in conventional boilers or in advanced combustion and gasification technologies, some of which can reduce coal-cleaning requirements. But, the use of low quality coal feed may increase the amount of waste generated at the power plant. The utilization of fine coal waste in conventional coal-fired power plants offer near-term opportunities for the reduction of fine coal waste disposed of in impoundments. However, the coal produced is more expensive than cleaned coal, as a result of capital and operating costs of additional equipment, and, in the case of coal water slurry, the additional cost of transportation. To compare technologies, the avoided costs of slurry impoundments must be included in a cost comparison.

Fluidized-bed combustion and gasification show promise for recovering the heat content of fine coal waste while avoiding some of the operational problems that limit use of coal fines in conventional pulverized coal-fired boilers. The combustion of fine coal waste in advanced combustion technologies, such as fluidized-bed combustion and gasification, is an alternative that also shows considerable long-term promise. Further research is needed on the use of fine coal waste slurries as feeds; incentives may be needed if these technologies are to be utilized widely for fine coal waste combustion. While coal combustion wastes from power plants are already being used for a number of purposes, the issue of the safe handling of coal combustion waste from these advanced combustion technologies should be studied further.

Both surface and underground methods are available for the disposal of coal slurry other than in impoundments. Alternative surface methods include incised ponds, slurry cells, combined refuse piles, and co-disposal of fine and coarse refuse. In many instances, these methods are influenced by topography, geology, and mining and coal preparation characteristics and are therefore site specific. The two primary methods for injecting fine coal refuse into underground mines are controlled flushing, where the underground workings are accessible, and blind or uncontrolled flushing, where the underground workings are abandoned or have caved in. A number of issues related to underground injection of slurry—such as adequate supply of water, surface ownership, permits, surface layout, and surface drainage—are independent of the method of slurry injection.

Although there are alternatives to disposing of coal waste in impoundments, no specific alternative can be recommended in all cases. Furthermore, the alternatives that have been identified are in varying stages of technological development and implementation. A factor limiting implementation to this point is the cost associated with the various alternatives. Additional research is needed to develop these alternatives further and to evaluate the economics of these processes. **The committee recommends that a screening study be conducted that (1) establishes ranges of costs applicable to alternative disposal options, (2) identifies best candidates for demonstration of alternative technologies for coal waste impoundments, and (3) identifies specific technologies for which research is warranted.** Input from MSHA and OSM regarding regulatory issues will be valuable to such a study. **The committee recommends that the use of economic incentives be explored as a way of encouraging the development and implementation of alternatives to slurry impoundments.** The development of incentives should be based on the full range of the portfolio of technologies as well as the economics of the technologies. The incentives should be linked directly to the reduction in slurry production of the utilization of slurry.

One way to reduce the volume of material in older slurry impoundments is to recover or remine the fine coal. Profitable remining may be possible if impoundments contain at least 1 million tons of in-situ slurry material, at an expected recovery rate of a marketable fine coal product of not less than 30 percent. The committee concludes that, as advances are made in the use of low value coal or coal water slurry, remining of slurry impoundments can be an attractive source of fuel.

ADDITIONAL RECOMMENDATIONS

In its deliberations, the committee identified several issues that cut across the elements of the statement of task. In addition, related areas warrant additional study, in the committee's view.

The committee notes that MSHA uses two systems to classify coal waste impoundments. One system classifies impoundments as high, medium, or low hazard, based on the magnitude of the potential consequences of a failure of the embankment structure—an embankment where people live and structures have been built immediately downstream would be classified as high hazard, regardless of the likelihood of failure of the embankment. A second system, which deals with basin failures, includes assessments of the proximity of underground mine workings, the potential for failure that would release slurry or water into those workings, and the potential impacts of a release. The second classification system comes closer to the standard definition of risk. Using different methodologies for classifying embankment and basins is inappropriate. **Therefore, the committee recommends that: (1) MSHA and OSM review activities related to risk assessment for existing impoundments (including both embankments and basins) to ensure that they are consistent and that they distinguish appropriately between hazard and consequence assessment in the methodologies adopted; and (2) MSHA and OSM establish a single, consistent system, which should be used to assign both embankments and basins to risk categories.** The ranking should be based on the combination of hazards and consequences, such as loss of life, cost, and environmental impact. This can be accomplished using qualitative risk assessment techniques.

A consistent risk assessment system would allow decisions on impoundments to be based on their relative risks. **The committee recommends that MSHA and OSM oversee a thorough assessment of potential mitigation measures for those impoundments that fall in the highest risk**

category and should determine which mitigation measures should be applied to reduce this risk to an acceptable level.

The committee also concludes that the design process for impoundments would be improved by a more formal risk analysis. Proposed new impoundments should also be assigned to risk categories, based on a combination of hazards and consequences, as was suggested for existing impoundments. **To maximize the potential for risk reduction, the committee recommends that all impoundment designs be accompanied by a risk analysis utilizing qualitative methods.** Examples of such methods include Potential Problem Analysis and Failure Modes and Effects Analysis.

The committee believes there is a limit to risk tolerance, for both existing and new impoundments. When risk is high, and when mitigation either through more reliable characterization or barrier construction is impossible, of limited precedent, or so expensive that it is infeasible, then a substantial change in operation of the impoundment is warranted. This may range from minimizing slurry fluidity to cessation of operation. Impoundments that fail risk-assessment criteria and where risk cannot be mitigated should be phased out or alternatives considered.

The committee heard repeatedly that the current review process for impoundment approval could take 2 years or more to complete. The committee believes that an efficient and coordinated regulatory review process would have substantial public benefit. A well-coordinated technical review process would protect health and safety of both miners and the public, and would foster protection of the environment. **Therefore, the committee recommends that the review process for both new permits and existing permits be overhauled to include the following elements:**

- **A formal joint review that would coordinate the currently fragmented and inefficient collection of reviews into a single process.**
- **Sufficient staff for engineering and other reviews in the agencies that participate in the joint process so that the time required to complete the review can be reduced significantly.**

The committee found that only very limited information was available concerning the quantities of trace elements in the slurry and the associated water. The committee heard repeatedly that citizens are concerned about ground and surface water quality and the impacts of impoundments on them. While a detailed review of the environmental impacts of coal waste impoundments is beyond the scope of this study, the committee identified this area as one needing further study. In addition, the character of the effluents will change with time as the coal and associated minerals including sulfides gradually oxidize. **The committee recommends that research be performed to identify the chemical constituents contained in the liquid and solid fractions of coal waste, and to characterize the hydrogeologic conditions around impoundments.** Information obtained from such research would have value for monitoring impoundments as well as for analyzing environmental impacts.

The committee also heard numerous comments indicating citizens' concern about emergency response and evacuation plans, which the operator of a coal waste impoundment must prepare to obtain a permit to build the impoundment. The committee recognizes the importance of the manner in which emergency and evacuation plans are developed, communicated with communities close to an impoundment, and coordinated with local emergency response authorities. Better communication among the companies operating impoundments, local emergency response authorities, and local citizens would allow the authorities and the citizens to understand the risks, the steps taken to mitigate them, and the appropriate responses in case of an accident.

SUMMARY

The conclusions and recommendations offered above reflect the committee's judgments concerning ways to improve the design process for coal waste impoundments, ways to improve mapping of mines and the characterization of sites of existing and future impoundments, and ways to improve the assessment and mitigation of risks associated with impoundments. The committee believes that implementation of those recommendations will substantially reduce the potential for uncontrolled release of coal slurry from impoundments, particularly through the mechanism of breakthrough into nearby underground mine workings. In addition, the committee believes that an appropriate way to balance alternatives for creating, handling, and disposing of wastes and to understand and mitigate the impacts of failure of any element of those systems is to view the designs of embankment and basins, as well as the entire process of handling and burning coal, as systems of interlinked components that operate together. The safe operation of these systems depends on effective engineering design, construction, and operation of facilities in addition to appropriate monitoring. With the recommended improvements in each of these areas, the potential for incidents like that at Inez, Kentucky, can be reduced.

References

Akers, D., Z.Zitron, R.Killmeyer, and G.Wen. 2001. Increasing the marketability of fine-sized coal. Presented at Coal Prep 2001, May 1–3, Lexington, KY.

Arey, D.G. 1997. Coal Technology Profiles. Carbondale, IL: Southern Illinois University Coal Research Center/Illinois Department of Commerce and Community Affairs.

ASCE [American Society of Civil Engineers]. 1999. Instrumentation of Embankment Dams and Levees, Technical Engineering and Design Guides No. 26. Reston, VA: ASCE.

ASTM [American Society for Testing and Materials]. 1999. Provisional Guide for Selecting Surface Geophysical Methods, ASTM PS 78–97, 10 pp. West Conshohocken, PA: ASTM.

Babcock, C.O., and V.E.Hooker. 1977. Results of Research to Develop Guidelines for Mining Near Surface and Underground Bodies of Water. U.S. Bureau of Mines Information Circular 8741, 24 pp.

Betoney, T.P. Jr. 1994. Mine Safety and Health Administration Accident Investigation Report (Surface Area of an Underground Coal Mine). Non-Injury Impoundment (Coal Refuse Pile —ID No 1211WV30054–00). Arkwright No. 1 Mine (ID No. 46– 01452), Consolidation Coal Company, Granville, Monongalia County, West Virginia, January 28, 5 pp.

Blakely, R.J. 1996. Potential Theory in Gravity and Magnetic Applications. Cambridge: Cambridge University Press, 441 pp.

BP Global 2001. Statistical Review of World Energy. Available at: http://www.bp.com/downloads/701/bp_global_stats.xls. Accessed September 14, 2001.

Branham, K.L., and D.W.Steeples. 1988. Cavity detection using high-resolution seismic reflection methods. Mining Engineering 40:115–119.

Brauner, G. 1973. Subsidence Due to Underground Mining, pt 1, Information Circular 8571, 56 pp.; pt. 2, Information Circular 8572. U.S. Bureau of Mines. 53 pp.

Breiner, S. 1973. Applications Manual for Portable Magnetometers. Sunnyvale, CA: GeoMetrics.

Brentrup, F.K. 1970. Seismische Vorfelderkundung zur Ortung tektonischer Stoerungen im Steinkohlebergbau. Glueckauf 106:933–938.

Bryar, T.R., C.J.Daughney, and R.J.Knight. 2000. Paramagnetic effects of iron (III) species on nuclear magnetic relaxation of fluid protons in porous media. Journal of Magnetic Resonance 142:74–85.

BTU Magazine. 1982. Coal Preparation—Quality Control in a Changing World. Special issue. Pittsburg, KS: McNally Pittsburgh, 47 pp.

Buchanan, D.J., R.Davis, P.J.Jackson, and P.M.Taylor. 1981. Fault location by channel waves seismology in United Kingdom coal seams. Geophysics 46:994–1002.

Cambridge, M. 2001. A review of tailings dams failures. Water Power and Dam Construction, May:40–42.

Cannon, W.C. 1981. Mine Safety and Health Administration Report of Investigation (Preparation Plant), Impoundment Accident (Failure of Refuse Pile). Preparation Plant (ID No. 15–11878), Eastover Mining Company, Brookside, Harlan County, Kentucky, December 18, 10 pp.

Chekan, G.J. 1985. Design of Bulkheads for Controlling Water in Underground Mines, U.S. Bureau of Mines Information Circular 9020, 36 pp.

Chircop, J., ed. 1999. Facts about Coal. Washington, DC: National Mining Association, 80pp.

Clayton, R.W., and R.H.Stolt. 1981. A born-WKBJ inversion method for acoustic structure beneath Southern California. Geophysical Research Letters 11:625–627.

Cook, J.C. 1965. Seismic mapping of underground cavities using reflection amplitudes. Geophysics 304:527–538.

Corwin, R.F. 1990. The self-potential method for environmental and engineering applications. Pp. 127–146 in Stan Ward, ed., Geotechnical and Environmental Geophysics. Tulsa, OK: Society for Exploration Geophysics.

Couch, Gordon R. 1991. Advanced Coal Cleaning Technology, IEACR/44. London: IEA Coal Research .

Couch, Gordon R. 1998. Adding Value to Coal Cleaning Wastes. London: IEA Coal Research.

Courier-Journal (Louisville). 1948. Fire destroys Norwood Hall, U.K. building: Loss put at $200,000; valuable records burn. November 13:P1C.

D'Appolonia Consulting Engineers. 1975. Engineering and Design Manual. Coal Refuse Disposal Facilities. Washington, DC: U.S. Department of the Interior, Mining Enforcement and Safety Administration.

Dana, P.H. 1999. The Geographer's Craft Project. Department of Geography, University of Colorado at Boulder. Available at: http://www.colorado.edu/geography/gcraft/notes/coordsys/coordsys_f.html. Accessed August 6, 2001.

Daniels, D.J. 1996. Surface-Penetrating Radar. London: Institute of Electrical Engineering, 320 pp.

Daniels, J.J., and W.S.Keys. 1990. Geophysical well logging for evaluating hazardous waste sites. Pp. 163–186 in Geotechnical and Environmental Geophysics, Stan Ward, ed. Tulsa, OK: Society for Exploration Geophysics.

Davies, M.P., and T.E.Martin. 2000. Upstream constructed tailings dams—a review of the basics. Pp. 3–15 in Tailings and Mine Waste. Rotterdam: Balkan.

Davies, M.R.C., A.G.Johnson, and K.P.Williams. 1998. Stabilised mixed colliery spoil in land reclamation. International Journal of Surface Mining, Reclamation, and Environment 12:1–4.

Davies, William E., James F.Bailey, and Donavan B.Kelly. 1972. West Virginia's Buffalo Creek Flood: A Study of the Hydrology and Engineering Geology, Circular 667. Reston, VA: U.S. Geological Survey.

Davis, J.L., and A.P.Annan. 1989. Ground-penetrating radar for high-resolution mapping of soil and rock stratigraphy. Geophysical Prospecting 37:531–552.

Dobecki, T.L., and P.R.Romig. 1985. Geotechnical and groundwater geophysics. Geophysics 50:2621–2636.

DOE [U.S. Department of Energy]. 1999. Integrated Gasification Combined Cycle. Washington, DC: Office of Fossil Energy, Federal Energy Technology Center, 12 p.

DOE. 2001. Fluidized Bed Coal Combustion. Available at: http://fossil.energy.gov/ coal_power/ fluidizedbed/index.shtml. Accessed August 27, 2001.

Domenico, S.N., and S.H.Danborn. 1987. Shear-wave technology in petroleum exploration—past, current, and future . Pp. 3–18 in Shear-Wave Exploration, S.H. Danborn and S.N.Domenico, eds. Tulsa, OK: Society of Exploration Geophysicists.

Downs, C.G., and J.Stocks. 1977. Environmental Impact of Mining. New York: John Wiley & Sons.

Dresen, L., and H.Reuter. 1994. Seismic coal exploration. Pp. 433 in In-Seam Seismics, vol. 16B, K.Helbig and S.Treitel, eds. New York: Pergamon.

Dunnicliff, J. 1988. Geotechnical Instrumentation for Monitoring Field Performance. New York: John Wiley and Sons.

Earth Resources Observation Systems (EROS) Data Center, Sioux Falls, SD. Available at: http:// edcwww.cr.usgs.gov/. Accessed September 10, 2001.

Ellis, R.G., and S.W.Oldenburg. 1994. The pole-pole 3-D DC-resistivity inverse problem: A conjugate-gradient approach. Geophysics Journal International 119:187– 194.

Energy Information Administration. 2001. U.S. Coal Supply and Demand: 2000 Review. Available at: http://www.eia.doe.gov/cneaf/coal/page/special/feature.html/ Accessed October 3, 2001.

EPA [U. S. Environmental Protection Agency]. 1999. Wastes from the Combustion of Fossil Fuels, vol. 2: Methods, Findings, and Recommendations. Report to Congress, EPA 530-R-99–010, March, 231 pp.

EPA. 2000. Regulatory Determination for Wastes from the Combustion of Fossil Fuels. Environmental Fact Sheet, EPA 530-F-00–025, May.

Fisher, W. 1971. Detection of Abandoned Underground Coal Mines by Geophysical Methods. Water Pollution Control Research Series, Project 14010EHN. Washington, DC: U.S. Environmental Protection Agency, 94 pp.

Fleischer, R.L., and A.Mogro-Campero. 1979. Radon enhancements in the earth: evidence for intermittent upflows? Geophysical Research Letters 6(5):361–364.

Frankel, A., C.Mueller, T.Barnhard, D.Perkins, E.V.Leyendecker, N.Dickman, S. Hanson, and M.Hopper. 1996. National Seismic Hazard Maps, June Documentation. U.S. Geological Survey Open-File Report 96–532. Reston, VA: U.S. Geological Survey, 110pp.

Franklin, John L., J.Steven Gardner, Samuel S.Johnson, and Kenneth P.Katen (Editorial Review Committee). 1997. Pp. 53–59 in Coal Mining Reference Book, 5th edition. Lexington: Kentucky Mining Institute.

Freme, F.L. 2001. U.S. Coal Supply and Demand: 2000 Review. Energy Information Administration , 8 pp.

Freme, F.L., and B.D. Hong. 2000. U.S. Coal Supply and Demand: 1999 Review. Energy Information Administration, 8 pp.

Garlanger, J.E., and N.F.Fuleihan. 1983. Reclamation options for clay settling areas. Pp. 32–58 in Proceedings of the Symposium on Reclamation and the Phosphate Industry. Barton, FL: Florida Institute of Phosphate Research.

Godson, R.H., and J.S.Watkins. 1968. Seismic resonance investigation of a near-surface cavity in Anchor Reservoir, Wyoming. Bulletin of the Association of Engineering Geologists 5 (1):27–36.

Goforth, T., and C.Hayward. 1992. Seismic reflection investigations of a bedrock surface buried under alluvium. Geophysics 57:1217–1227.

Goldbach, G., and M.Tanca. 2001. Firing coal washing wastes in a Fi Circ™ steam generator, Redbank Power Project. Presentation by Alstom Power at the International Fluid Bed Conference, Reno, NV, May, 16 pp.

Graedel, T.E., and B.R.Allenby. 1995. Industrial Ecology. Upper Saddle River, NJ: Prentice Hall.

Gray, R.E., J.C.Gamble, R.J.McLaren, D.J.Rogers. 1974. State-of-the-Art of Subsidence Control, Final Report (December), No. ARC-73-111–2550, Appalachian Regional Commission, 265 pp.

Greenhalgh, S.A., D.Burns, and I.Mason. 1986. A cross-hole and face-to-borehole in-seam seismic experiment at Invincible Colliery, Australia. Geophysical Prospecting 34:30–55.

Gritto, R., and L.Dresen. 1992. Seismic modeling of seam waves excited by energy transmission into a seam. Geophysical Prospecting 40:671–699.

Harlow, G.E., and G.D.Le Cain. 1991. Hydraulic Characteristics of, and Ground-Water Flow in, Coal-Bearing Rocks of Southwestern Virginia, Open-File Report 91–250. Reston, VA: U.S. Geological Survey.

Harrison, C.D., and D.J.Akers. 1997. Coal Fines—Resource of the Future. Conference Proceedings. Coal Liquefaction and Solid Fuels Contractors Review Conference, Federal Technical Center Publications, Pittsburgh, PA. Available at: http://www.netl.doe.gov/publications/proceedings/97/97cl/harrison.pdf. Accessed August 6, 2001.

Hasbrouck, W.P. 1987. Hammer impact, shear-wave studies. Pp. 97–121 in Shear-Wave Exploration, S.H.Danborn and S.N.Domenico, eds. Tulsa, OK: Society of Exploration Geophysicists.

Hasbrouck, W.P. 1991. Four shallow-depth, shear-wave feasibility studies. Geophysics 56:1875–1885.

Hasbrouck, W.P., and N.Padget. 1982. Use of shear wave seismics in evaluation of strippable coal resources. Utah Geological and Mineral Survey Bulletin 118:203– 210.

Hatchett, E.B. Jr. 2001. Kentucky's Management of Nonpoint Source Water Pollution. August Performance Audit, Commonwealth of Kentucky Auditor of Public Accounts, 43 pp.

Hearst, J.R., P.H.Nelson, and F.L.Paillet. 2000. Well Logging for Physical Properties, 2nd edition. New York: John Wiley and Sons, 483 pp.

Hebden, D., and H.J.F.Stroud. 1981. Coal gasification processes. Pp. 1599–1752 in Chemistry of Coal Utilization, 2nd suppl. vol., M.A.Elliott, ed. Washington, DC: National Academy of Sciences.

Heirendt, K.A. 1988. An analysis of ^{222}Rn soil gas concentrations in the Serpent Mound area, Southwestern Ohio. M.S. Thesis, University of Akron, Akron, Ohio, 86 pp.

Howard, K.W.F. 1990. Geophysical well logging methods for detection and characterization of fractures in hard rocks. Pp. 287–308 in Geotechnical and Environmental Geophysics, Stan Ward, ed. Tulsa, OK: Society for Exploration Geophysics.

Humphreys, G., R.W.Clayton, and B.H.Hage. 1984. A Tomographic Image of Mantle Exploration, 3rd edition. New York: John Wiley and Sons, 538 pp.

Hunter, J.A., S.E.Pullan, R.A.Burns, R.M.Gagne, and R.L.Good. 1984. Shallow seismic reflection mapping of the overburden-bedrock interface with the engineering seismograph—some simple techniques. Geophysics 49:1381–1385.

Hutchinson, I.P.G., and R.D.Ellison, eds. 1992. Mine Waste Management. Sacramento: California Mining Association/ Lewis Publishers, Inc.

Hynes, Norman J. 1995. Nontechnical Guide to Petroleum Geology, Exploration, Drilling, and Production. Tulsa, OK: Penn Well Publishing, 536 pp.

ICOLD [International Commission on Large Dams], Committee on Seismic Aspects of Dam Design. 1989a. Selecting Seismic Parameters for Large Dams. Bulletin 72. Paris, 73 pp.

ICOLD, Committee on Mining and Industrial Tailings Dams. 1989b Tailings Dams Safety—Guidelines. Bulletin 74. Paris, 107 pp.

ICOLD, Committee on Mining and Industrial Tailings Dams. 1994. Tailings Dams— Design of Drainage. Bulletin 97. Paris, 119 pp.

ICOLD, Committee on Mining and Industrial Tailings Dams. 1995a. Tailings Dams and Seismicity —Review and Recommendations. Bulletin 98. Paris, 60 pp.

ICOLD, Committee on Mining and Industrial Tailings Dams. 1995b. Tailings Dams. Transport, Placement, Decantation—Review and Recommendations. Bulletin 101. Paris, 97 pp.

ICOLD, Committee on Mining and Industrial Tailings Dams. 1996. A Guide to Tailings Dams and Impoundments—Design, Construction, Use, and Rehabilitation. Bulletin 97. Paris: ICOLD and UNEP, 239 pp.

ICOLD, Committee on Tailings Dams and Waste Lagoons. 2001. Tailings Dams—Risk of Dangerous Occurrences—Lessons Learnt from Practical Experiences. Bulletin 21. Paris, 144pp.

lyengar, R.K., and K.Ramakrishnan. 1992. Coal gasification and utilisation. Pp. 25–38 in International Conference on Environmentally Sound Technologies: Policy Issues and Technological Options, 15–18 January 1998, Madras, India. New York: United Nations Centre for Science and Technology for Development.

Jelinski, L.W., T.E.Graedel, R.A.Laudise, D.W.McCall, and C.K.N.Patel. 1992. Industrial ecology: concepts and approaches. Proceedings of the National Academy of Sciences 89(3):793–797.

Johnson, J.L. 1981. Fundamentals of coal gasification. Pp. 1491–1598 in Chemistry of Coal Utilization, 2nd suppl. vol., M.A.Elliott, ed. Washington, DC: National Academy of Sciences.

Johnson, T.L. 1999. Design of Erosion Protection for Long-Term Stabilization, Draft, NUREG-1623. Washington, DC: U.S. Nuclear Regulatory Commission.

Kipp, J. A, and J.S.Dinger. 1987. Stress-relief fracture control of groundwater movement in the Appalachian Plateau. Pp. 423–438 in Focus Conference on Eastern Regional Ground-Water Issues, National Water Well Association, July 15, Burlington, VT.

Kipp, J. A, F.W.Lawrence, and J.S.Dinger. 1983. A conceptual model of ground-water flow in the Eastern Kentucky Coal Field: Symposium on Surface Mining, Hydrology, Sedimentation, and Reclamation, University of Kentucky, November 27-December 2, Lexington, KY.

Knight, R., T.Bryar, and C.Daughney. 1999. Laboratory studies to assess the use of nuclear magnetic resonance for near-surface applications . Exposition Abstract #NSG2.3. 1999 International Exposition and 69th Annual Meeting, Houston, Texas. Tulsa, OK: Society of Exploration Geophysics.

Lankston, R.W., and M.M.Lankston. 1986. Obtaining multilayer reciprocal times through phantoming. Geophysics 51:45–49.

Leonard, J. 1991. Coal Preparation, 5th edition. Littleton, CO: Society for Mining, Metallurgy, and Exploration.

Leps, T.M. 1970. Review of the shearing strength of rockfill. Journal of the Soil Mechanics and Foundation Engineering Division 4(July):1159–1170.

Lexington Herald Leader. 1948. UK building is destroyed by early-morning blaze. November 13: P1C.

Lindsey, C.G., L.W.Long, and C.W.Begej. 1982. Long-Term Survivability of Riprap for Armoring Uranium Mill Tailings and Covers: A Literature Review, NUREG/CR-2642. Washington, DC: U.S. Nuclear Regulatory Commission.

MAC [Mining Association of Canada]. 1998. A Guide to the Management of Tailings Facilities. Ottawa: MAC.

Marsal, R.J. 1973. Mechanical properties of rockfill. In Embankment Dam Engineering, R.C.Hirschfeld and S.J.Poulos, eds. New York: John Wiley and Sons.

Martin, T.E., and M.P.Davies. 2000. Trends in the stewardship of tailings dams. Pp. 393–407 in Tailings and Mine Waste '00, Fort Collins, CO. Rotterdam: Balkema.

Mason, I. 1981. Algebraic reconstruction of a two-dimensional velocity inhomogeneity in the High Hazle seam of Thoresby colliery. Geophysics 46:298–308.

Masudi, H., and S.Samudrala. 1996. Combustion analysis of coal-water slurry fuel prepared from plant coal and recovered coal fines. Conference Proceedings: Joint Power and Fuel Systems Contractor Review Meeting, July 9–11. Available at: http://www.fetc.doe.gov/publications/proceedings/96/96jpfs/jpfs_pdf/cws.pdf. Accessed August 6, 2001

McMurray, J. 1984. Organic Chemistry. Monterrey, CA: Brooks/Cole Publishing.

McNeill, J.D. 1990. Use of electromagnetic methods for groundwater studies. Pp. 191– 218 in Geotechnical and Environmental Geophysics, Stan Ward, ed. Tulsa, OK: Society for Exploration Geophysics.

Michalek, Stanley J., George H.Gardner, and Kelvin K.Wu. 1996. Accidental Releases of Slurry and Water from Coal Impoundments through Abandoned Underground Coal Mines. Pittsburgh, PA: Mine Safety and Health Administration, Safety and Health Technology Center.

Miller, R.D., and D.W.Steeples. 1991. Detecting voids in a 0.6-m coal seam, 7 m deep, using seismic reflection . Geoexploration 28:109–119.

Miller, R.D., and D.W.Steeples. 1995. Applications of shallow high-resolution seismic reflection to various mining operations. Mining Engineering 47:355–361.

Minns, S.A. 1993. Conceptual Model of Local and Regional Ground-Water Flow in the Eastern Kentucky Coalfield, Ph.D. Kentucky Geological Survey Thesis Series 6, Series XI, 194pp.

Misquitta, N.J. 1989. Enhanced ^{222}Rn level over an abandoned underground coal mine. M.S. Thesis, University of Akron, Akron, Ohio, 72 pp.

Moebs, N.N., and G.P.Sames. (1989). Leakage across a Bituminous Coal Mine Barrier, Report of Investigations 9280. Washington, DC: U.S. Bureau of Mines, 26 pp.

Morgenstern, N.R. 1995. Managing Risk in Geotechnical Engineering, Third Casagrande Lecture. Pp. 102–126, vol. 4 in Proceedings of the 10th Pan American Conference on Soil Mechanics and Foundation Engineering, Guadalajara, Mexico.

MSHA [Mine Safety and Health Administration]. 1974. Design Guidelines for Coal Refuse Piles and Water, Sediment, or Slurry Impoundments, and Impounding Structures. Informational Report 1109. U.S. Department of Labor, 29 pp.

MSHA. 1983. Design Guidelines for Coal Refuse Piles and Water, Sediment, or Slurry Impoundments, and Impounding Structures. Amendment to Informational Report 1109. U.S. Department of Labor , 6 pp.

MSHA. 1997. Procedure instruction letter I99-V-3. December 1.

Nabighian, M.N., ed. 1988. Electromagnetic Methods in Applied Geophysics, vol. 1, 1st edition. Tulsa, OK: Society of Exploration Geophysicists.

Nabighian, M.N., ed. 1991. Electromagnetic Methods in Applied Geophysics, vol. 1, 2nd edition. Tulsa, OK: Society of Exploration Geophysicists.

National Geologic Map Database: Geologic Names Lexicon. Available at: http://ngmdb.usgs.gov/. Accessed September 12, 2001.

Nielson, D.L., Cui Linpei, and Stanley H.Ward. 1990. Gamma-ray spectrometry and radon emanometry in environmental geophysics. Pp. 219–250 in Geotechnical and Environmental Geophysics, Stan Ward, ed. Tulsa, OK: Society for Exploration Geophysics.

NRC [National Research Council]. 1975. Underground Disposal of Coal Mine Wastes: A Report to the National Science Foundation Study Committee to Assess the Feasibility of Returning Underground Coal Mine Wastes to the Mined-Out Areas. Washington, DC: National Academy Press, 172 pp.

NRC. 1994. Drilling and Excavation Technologies for the Future. Washington, DC: National Academy Press, 161 pp.

NRC. 1995. Earth Observations from Space: History, Promise, and Reality. Washington, DC: National Academy Press, 310 pp.

NRC. 2000a. Seeing into the Earth—Noninvasive Characterization of the Shallow Subsurface for Environmental and Engineering Applications. Washington, DC: National Academy Press, 129 pp.

NRC. 2000b. Vision 21: Fossil Fuel Options for the Future. Washington, DC: National Academy Press. 144pp.

NRC. 2001. Evolutionary and Revolutionary Technologies for Mining. Washington, DC: National Academy Press, 159 pp.

Oder, R.R., E.D.Brander, and R.E.Jamison. forthcoming. Concentration of trace metals inside coal pulverizers. In Proceedings of the 11th International Conference on Coal Science, San Francisco, CA.

Office of Surface Mines. Mine Map Repository (National), Pittsburgh, Pennsylvania. Available at: http://mmr.osmre.gov. Accessed September 10, 2001.

Office of Technology Assessment. 1979. Direct Use of Coal: Prospects and Problems of Production and Combustion, OTA-E-86. Washington, DC, 411 pp.

Osborne, D.G. 1988. Coal Preparation Technology, vol. 1. London: Graham and Trotman.

Owens, H.L. 1977. Mining Enforcement and Safety Administration Memorandum to Robert Fujimoto, Chief Mine Waste Branch, Denver Technical Support Center. Failure of the Fine Coal Refuse Impoundment, Pond Fork Preparation Plant, Island Creek Coal Company, Bob White, Boone County, West Virginia, ID No. 1211WV40009–02, 8pp.

Owens, H.L. 1987. Mine Safety and Health Administration Report of Investigation (Preparation Plant). Impoundment Principal Spillway Pipe Failure. Montcoal Preparation Plant (ID No. 46–03430), Peabody Coal Company. Montcoal, Raleigh County, West Virginia. April 27.

Palmer, D. 1980. The Generalized Reciprocal Method of Seismic Refraction Interpretation. Tulsa, OK: Society of Exploration Geophysicists.

Park, C.B., R.D.Miller, and J.Xia. 1999. Multichannel analysis of surface waves. Geophysics 64:800–808.

Peck, R.B. 1969. Advantages and limitations of the observational method in applied soil mechanics, Ninth Rankine Lecture. Geotechnique 19(2): 171–187.

Penman, A.D. M., K.R.Saxena, and V.M.Sharma. 1999. Instrumentation, Monitoring and Surveillance-Embankment Dams. Brookfield, WI: A.A.Balkema.

Pennsylvania Department of Environmental Protection. 1998. Coal Mine Drainage Prediction and Pollution Prevention in Pennsylvania. Available at: http://www.dep.state.pa.us/dep/deputate/minres/districts/cmdp/main.htm. Accessed September 12, 2001.

Pennsylvania Department of Environmental Protection. 1999. Engineering Manual for Mining Operations. Pennsylvania Bureaus of Mining and Reclamation and District Mining Operations, Document No. 563–0300–101, January.

Peterson, D.J., T.LaTourrette, and J.T.Bartis. 2001. New Forces at Work in Mining: Industry Views of Critical Technologies. Arlington, VA: Rand Science and Technology Policy Institute, 92 pp.

Raeder, D., W.Schott, L.Dresen, and H.Reuter. 1985. Calculation of dispersion curves and amplitude-depth distribution of Love channel waves in horizontally layered media. Geophysical Prospecting 33:800–816.

Rechtien, R.D., and D.M.Stewart. 1975. A seismic investigation over a near-surface cavern. Geoexploration 13:235–246.

Rice, C.L., E.G.Sable, G.R.Dever, and T.M.Kehn. 1979. The Mississippian and Pennsylvanian (Carboniferous) Systems in the United States-Kentucky. United States Geological Survey Professional Paper, 1110-F. Reston, VA: U.S. Geological Survey, 32 pp.

Roberts, R.L., W.J.Hinze, and D.I.Leap. 1990a. Data enhancement procedures on magnetic data from landfill investigations. Pp. 261–266 in Geotechnical and Environmental Geophysics, vol. 2: Environmental and Groundwater, S.H.Ward, ed. Tulsa, OK: Society of Exploration Geophysics.

Roberts, R.L., W.J.Hinze, and D.I.Leap. 1990b. Application of the gravity method to the investigation of a landfill in glaciated midcontinent, U.S.A. Pp. 253–260 in Geotechnical and Environmental Geophysics, vol. 2: Environmental and Groundwater S.H.Ward, ed. Tulsa, OK: Society of Exploration Geophysics.

Robinson, E.S., and C.Coruh. 1988. Basic Exploration Geophysics. New York: John Wiley and Sons, 562 pp.

Rousaki, K., and G.Couch. 2000. Advanced Clean Coal Technologies and Low Value Coals, CCC/39. London: IEA Coal Research, 76 pp.

Schiffman, R.L., S.G.Vick, and R.E.Gibson. 1988. Behavior and properties of hydraulic fills. Pp. 166–202 in Hydraulic Fill Structures, Geotechnical Special Publication No. 21, D.J.A.Van Zyl and S.G.Vick, eds. Reston, VA: American Society of Civil Engineers.

Schwaetzer, T. 1965. Geophysical studies of the continuity of coal seams. International Journal of Rock Mechanics and Mining Science 1965:167–196.

Seed, H.B. 1979. Considerations in the earthquake-resistant design of earth and rockfill dams. Geotechnique 29(3):215–263.

Semenov, A.G., A.I.Burshtein, A.Yu.Pusep, and M.D.Schirov. 1988. A device for measurement of underground mineral parameters: U.S.S.R. Patent 1,079,063.

Shackleford, Raymond. 2001. Underground Surveying—a Primer in Coal Mine Surveying and Mapping. Available at: http://www.pobonline.com/CDA/ArticleInformation/features/BNP_Features_Item/0,2338,11829,00.html. Accessed September 10,2001.

Shadbolt, C.H. 1977. Mining subsidence—historical review and state of the art. Pp. 1–54 in Proceedings, Conference on Large Ground Movements and Structures, University of Wales, Institute of Science and Technology, Cardiff, Wales, J.D.Geddes, ed. New York: John Wiley and Sons.

Sharma, P.V. 1997. Environmental and Engineering Geophysics. Cambridge: Cambridge University Press, 475 pp.

Shirov, M., A.Legchenko, and G.Creer. 1991. A new direct non-invasive ground water detection technology for Australia. P. 17 in Exploration in a Changing Environment: Abstracts of the Australian Society of Exploration Geophysicists 8th Conference and Exhibition and the Geological Society of Australia Exploration Symposium, W. Jamieson and T. Pippett, eds. Brisbane, Queensland: Geological Society of Australia.

Singh, M.M. 1992. Mine subsidence. Pp. 938–971 in SME Mining Engineering Handbook, 2nd edition. Littleton, CO: Society for Mining, Metallurgy, and Exploration.

Singh, M.M., and F.S.Kendorski. 1981. Strata disturbance prediction for mining beneath surface water and waste impoundments. Pp. 76–89 in Proceedings of the First Conference on Ground Control in Mining, West Virginia University, Morgantown.

Snyder, J.P. 1987 [reprinted with changes 1997]. Map Projections—a Working Manual, U.S. Geological Survey Professional Paper 1385. Reston, VA: U.S. Geological Survey, 383 pp.

Stach, E. 1982. Coal Petrology. Berlin: Gebruder Borntraeger, 535 pp.

Steeples, D.W., and R.D.Miller. 1990. Seismic-reflection methods applied to engineering, environmental, and ground-water problems. Pp. 1–30 in Geotechnical and Environmental Geophysics, Stan Ward, ed. Tulsa, OK: Society for Exploration Geophysics.

Steeples, D.W., G.S.Baker, and C.Schmeissner. 1999. Toward the autojuggie: Planting 72 geophones in 2 seconds. Geophysical Research Letters 26:1085–1088.

Stewart, B., and G.Robinson. 1994. Mine Safety and Health Administration Accident Investigation Report (Preparation Report). Non-Injury Inundation Report. Martin County Coal Corporation, Preparation Plant (ID No. 15–03752), 1-C Mine (ID No. 15–03752), Davella, Martin County, Kentucky, May 23, 9 pp.

Stokoe, K.H., S.G.Wright, S.A.Barg, and J.M.Roesact. 1994. Characterization of geotechnical sites by SASW method. Pp. 15–25 in Geophysical Characterization of Sites, R.D.Woods, ed. Oxford: B.H. Publishing.

Sweigard, R. 1992. Reclamation. Pp. 1181–1197 in H.L.Hartman, ed. Mining Engineering Handbook, 2nd edition. Littleton, CO: Society for Mining, Metallurgy, and Exploration.

Syngle, D.V., and B.T.Sinn. 1991. CFB boiler fires waste coal, achieves high availability. Power (April):91–96.

Thacker, B.K. 2000. Perimeter embankments to increase capacity, mitigate subsidence impacts, and abandon coal refuse disposal impoundments. Pp. 143–154 in Tailings

Dams 2000 Proceedings, Association of State Dam Safety Officials, U.S. Committee on Large Dams.

Thacker, B.K., C.R.Ullrich, J.G.Athanasakes, and G.Smith. 1988. Evaluation of a coal refuse impoundment built by the upstream method. Pp.730–749 in Hydraulic Fill Structures, Geotechnical Special Publication No. 21, D.J.A.Van Zyl and S.G. Vick , eds. Reston, VA: American Society of Civil Engineers.

Tillard, S., and J.C.Dubois. 1995. Analysis of GPR data: Wave propagation velocity determination. Journal of Applied Geophysics 33:77–91.

Toothman, F.R. 1977. Correlation Chart, Major Mineable Bituminous Coal Seams, Appalachian Coal Fields. Huntington, WV: Vandalia Book Co.

U.S. Army Corps of Engineers. 1982. Engineering and Design Stability for Earth and Rock-Fill Dams, EM 1110–2–1902. Washington, DC: Government Printing Office.

U.S. Army Corps of Engineers. 1994. Earth and Rock-fill Dams—General Design and Construction Considerations, Manual 1110–2–2300. Washington, DC.

U.S. Army Corps of Engineers. 1995. Instrumentation of Embankment Dams and Levees, Manual 1110–2–1908. Washington, DC.

U.S. Bureau of Reclamation. 1987. Embankment Instrumentation Manual. Denver, CO.

U.S. Bureau of Reclamation. 1988. Downstream Hazard Classification Guidelines. Assistant Commissioner—Engineering and Research Technical Memorandum No. 11. Denver, CO, 14pp.

U.S. Department of Agriculture, 1976. Earth Dams and Reservoirs. Soil Conservation Service Technical Release No. 60, June, 53 pp.

USCOLD [U.S. Committee on Large Dams], Committee on the Monitoring of Dams and Their Foundations. 1993. General Guidance and Current U.S. Practice in Automated Performance Monitoring of Dams of the United States. Denver.

USCOLD. 1994. Tailings Dam Incidents. Denver.

Vick, S.G. 1990. Planning, Design, and Analysis of Tailings Dams. Vancouver, BC: BiTech Publishers, 369 pp.

Vick, S.G. 2000. Tailings Dam Safety—Implications for the dam safety community. Pp. 1–20 in Proceedings of Tailings Dams 2000. Las Vegas, NV: Association of State Dam Safety Officials.

Wagh, A.S., and P.Desai. 1987. Bauxite Tailings. Kingston, Jamaica: Jamaica Bauxite Institute/ University of the West Indies.

Ward, S.H. 1990. Geotechnical and Environmental Geophysics, vol. 1–3. Tulsa, OK: Society of Exploration Geophysicists.

Ward, S.H., B.K.Sternberg, D.J.LaBrecque, and M.M.Poulton. 1995. Recommendations for IP research. The Leading Edge 14:243–247.

Waters, Kenneth H. 1987. Reflection seismology: A tool for energy resource reflection data. Geophysics 46:1559–1567.

Watkins, J.S., R.H.Godson, and K.Watson. 1967. Seismic Detection of Near-Surface Cavities. U.S. Geological Survey, Professional Paper 599-A. Reston, VA: U.S. Geological Survey, 12 pp.

Watson, K., and D.H.Knepper, eds. 1994. Airborne Remote Sensing for Geology and Environment —Present and Future. U.S. Geological Survey Bulletin 1926. Reston, VA: U.S. Geological Survey, 43 pp.

Weinheimer, R.L. 1983. Vertical Sequence and Diagenesis in Breathitt Sandstone of Eastern Kentucky. M.S.Thesis, University of Cincinnati, Cincinnati, Ohio, 147 pp.

Whitehead, J. 1997. Briquetting Coal to Enhance Value. Paper presented at the Conference on Adding Value to Coal, June 25–26, Rotterdam, Netherlands. London: Coal Transactions, 6 pp.

Williams, D., M.Gowan, and P.Keefer. 1995. Practical co-disposal deposition. In Proceedings of the Seventh Australian Coal Preparation Conference, J.Smitham, ed. Australian Coal Preparation Society.

Wilson, S.D., and R.J.Marsal. 1979. Current Trends in Design and Construction of Embankment Dams, prepared for International Commission on Large Dams and Geotechnical Division of the American Society of Civil Engineers . New York: American Society of Civil Engineers.

Wunsch, D.R. 1993. Ground-water Geochemistry and Its Relationship to the Flow System at an Unmined Site in the Eastern Kentucky Coal Field, Kentucky Geological Survey Thesis Series 5, Series XL

Wyrick, G.C., and J.W.Borchers. 1981. Hydrologic effects of stress relief fracturing on an Appalachian valley, U.S. Geological Survey Water-Supply Paper 2177. Reston, VA: U.S. Geological Survey, 51 pp.

Xia, J., R.D.Miller, and C.B.Park. 1999. Estimation of near-surface shear-wave velocity by inversion of Rayleigh waves. Geophysics 64:691–700.

Yule, D.E., M.K.Sharp, and D.K.Butler. 1998. Microgravity investigations of foundation conditions. Geophysics 63:95–103.

Zohdy, A.A.R. 1989. A new method for the automatic interpretation of Schlumberger and Wenner sounding curves. Geophysics 54:245–253.

APPENDIXES

Appendix A

Biographical Sketches of Committee Members

FRANKLIN M.ORR, JR., *chair,* is the Beal Professor and Dean of the School of Earth Sciences at Stanford University. His research interests include multicomponent fluid phase equilibrium and its interactions with multiphase flow in porous media. Previously he served as chair of the Petroleum Engineering Department at Stanford University and held positions at the New Mexico Institute of Mining and Technology, Shell Development Company, and the U.S. Environmental Protection Agency. He is a member of the Society of Petroleum Engineers, the American Institute of Chemical Engineers, and the Society of Industrial and Applied Mathematics. He is vice-chair of the board of directors of the Monterey Bay Aquarium Research Institute and member of the board of directors of the David and Lucile Packard Foundation Fellowships in Science and Engineering. He is also a member of the National Academy of Engineering and has served as chair on the Panel for Review of the Energy Resources Program of the U.S. Geological Survey.

GARY A.DAVIS is the founder and director of the University of Tennessee Center for Clean Products and Clean Technologies, a senior fellow at the University of Tennessee Energy, Environment, and Resources Center, and an adjunct professor of environmental law at the University of Tennessee. He holds a B.S. in chemical engineering from the University of Cincinnati and a J.D. from the University of Tennessee. He has conducted research on the life-cycle environmental impacts of products, substitutes for polluting products, and policies to encourage the use of cleaner products and processes. Mr. Davis has been working on technical and policy issues related to pollution prevention for more than 20 years. He has published numerous books and articles on a variety of environmental issues. He was previously with the California Governor's Office, where he worked on hazardous waste and hazardous substance policy. He has also practiced environmental law for 17 years.

BARBARA A.FILAS, P.E., vice president, Mining and Environment, Knight Piesold Consulting, is a mining engineer with more than 20 years of experience in surface and underground mine operations, engineering, and regulatory support for coal, metals, and industrial mineral mining projects. Knight Piesold is one of the top dam design consulting firms in the world. Previously Ms. Filas was an engineer with Atlas Minerals, Summit Minerals, Monterey Coal Company, and U.S. Steel Corporation with expertise in waste containment facility design, reclamation plans and surety estimates, environmental site and compliance audits, and storm water and sedimentation control designs. Ms. Filas holds a B.S. in mining engineering from the University of Arizona. She is a member of several professional organizations such as the Society for Mining, Metallurgy, and Exploration, Inc. (having served as chair of its Environmental Division and on its Board of Directors) and the National Society of Professional Engineers, and currently serves as a program reviewer for the Accreditation Board of Engineering and Technology. She has published numerous articles on environmental aspects of new project development, environmental controls, and mine waste disposal issues and mine closure and reclamation.

C.DAVID HENRY is vice president of operations, Beard Technologies, Inc. There his focus is to develop coal recovery operations through the utilization of advanced technologies and to produce a high-grade fine coal product to be sold in the general coal market. Previously, he held engineering positions at C.D.H. Consulting and Mineral Development Corporation. His area of expertise includes recovery and reclamation of coal slurry impoundments, testing and analysis of coal slurry samples, slurry pond reclamation design, and coal preparation. He has designed a sampling system to extract slurry materials from impoundment structures and a dredging unit.

NORBERT R.MORGENSTERN is a university professor of civil engineering (emeritus), University of Alberta, and an internationally recognized authority in the field of geotechnical engineering. He was key to the development of one of the leading geotechnical schools, bringing about the foundation of modern practice in permafrost engineering and slope design. Professor Morgenstern has received numerous awards, among them: the Walter Huber Engineering Research Prize from the American Society of Civil Engineers, the Canadian Geotechnical Society Prize, including the Legget Award, the Alberta Order of Excellence, and the Order of Canada. His publications number approximately 300, dealing with foundation engineering, environmental issues, mine abandonment, soil and rock mechanics, embankment dams, and arctic soils and excavation. He is affiliated with several professional associations and has served on various committees, including the International Society for Soil Mechanics and Foundation engineering (past president), the Engineering Institute of Canada (past president), the Canadian Geoscience Council (past president), the Canadian Geotechnical Society, the Association of Engineering Geologists, and the Geological Society of London. Professor Morgenstern has consulted on projects in 35 countries and has assisted in technology transfer to developing countries through the United Nations and other agencies. He is also a member of the National Academy of Engineering and a fellow of the National Academies of Engineering in Canada, the United Kingdom, and India.

DAVID A.NEWMAN, P.E., P.G., is president of Appalachian Mining and Engineering, Inc./Geolab and president of Newman Engineering, PSC. From 1984 to 1988, he was an assistant professor of mining engineering at the University of Kentucky. He holds a Ph.D. in mining engineering from Pennsylvania State University. Dr. Newman's areas of expertise include rock and soil mechanics, geotechnical engineering, subsidence prediction and abatement, slope stability, analysis of refuse impoundment stability, underground mine stability, and evaluation of underground mine workings beneath coal refuse impoundments. He holds patent disclosures for sampling devices and mining equipment. He has directed a number of projects on slope stability, focusing on computer-based analysis of soil slopes and impoundments; operational and regulatory considerations for a slurry impoundment; and slurry reclamation. Dr. Newman's memberships in professional societies include the Society for Mining, Metallurgy, and Exploration, Inc., serving on the Rock Mechanics Award Committee, as chair and the Professional Engineering Examination Committee; and the Acid Mine Drainage Committee of the National Coal Association.

RAJA V.RAMANI, P.E., holds the Anne B. and George H. Jr. Deike Chair in mining engineering at the Pennsylvania State University where he has been on the faculty since 1970 and is a professor of mining and geo-environmental engineering. His research activities include 6 years of experience in the coal mining industry, flow mechanisms of air, gas, and dust in mining environs, innovative mining methods, and health, safety, productivity, and environmental issues in the mineral industry. He has published more than 200 research papers, contributed to 25 books, and edited the proceedings of 15 national and international symposiums. Dr. Ramani has been a consultant to the United Nations and the World Bank and has received numerous awards from academic and technical and professional societies. He was the 1995 president of the Society for Mining, Metallurgy, and Exploration, Inc. He served on the U.S. Department of Health and Human Services' Mine Health Research Advisory Committee (1991–1998). He was the chair of the U.S. National Academy of Sciences' Committee on Post Disaster Survival and Rescue (1979–1981), and served on the NAS Committee on Mining Technologies (2000–2001) and the Health Research Panel of the NAS Committee on the Research programs of the U.S. Bureau of Mines (1994). He was a member of the Department of the Interior's Advisory Board to the Director of U.S. Bureau of Mines (1995), and a member of the Secretary of Labor's Advisory Committee on the Elimination of Coal Workers' Pneumoconiosis (1995–1996).

ROBERT L.SCHUSTER, P.E., P.G., is an engineering geology and geotechnical engineering consultant. He retired from the U.S. Geological Survey in 1995 but continues to serve that agency as a scientist emeritus. His research interests include slope failure, engineering geologic aspects of natural, water-storage, and tailings dams, and geologic hazards mitigation. He holds a Ph.D. from Purdue University in civil engineering. He has served as professor of civil engineering at the University of Colorado, professor and head of the Department of Civil Engineering, University of Idaho, and chief of the Engineering Geology Branch, U.S. Geological Survey. Among other honors, Dr. Schuster has been awarded the Distinguished Service Award of the U.S. Department of the Interior, the International Meritorious Service Award of The Japan Landslide Society, a NATO Senior Fellowship in Science to the University of London, and a Senior Fulbright Scholarship to MacQuarie University, Australia. He is a member or fellow of numerous professional societies and has served on several NRC boards and committees. Dr. Schuster has written or edited more than 250 papers, texts, and reports on topics in engineering geology and geotechnical engineering.

MADAN M.SINGH, P.E., is president of Engineers International, Inc. He has held research positions at the Pennsylvania State University and IIT Research Institute. Dr. Singh's research interests and expertise encompass diverse aspects of rock mechanics, mining, hydrogeology, and geotechnical engineering. He developed a graduate-level course in mine subsidence engineering at the Pennsylvania State University and acted as advisor during the drafting of subsidence-control legislation in the Commonwealth of Pennsylvania. He has a Ph.D. in mining engineering from the Pennsylvania State University. Dr. Singh has served in several capacities for professional societies, including national director of the American Consulting Engineers

Council, president of the Consulting Engineers Council of Illinois, member of the Board of Directors of the Society for Mining, Metallurgy, and Exploration, Inc. (SME), chair of the SME Coal Division, and chair of the American Society for Testing and Materials subcommittee on rock strength. He has authored more than 100 technical papers, in addition to serving as chapter author on mine subsidence in the SME Mining Engineering Handbook (also associate editor) and Mining Environmental Handbook and, as editor of the "Legislative Update" for the Hazardous Waste Action Coalition. Dr. Singh has served on two NAS/NRC committees, the U.S. National Committee on Rock Mechanics (1977–1980) and the U.S. National Committee on Tunneling Technology (1974–1976). He was named a Centennial Fellow by the College of Earth and Mineral Sciences (1996) and honored with the Robert Stefanko Distinguished Achievement Award by the Department of Energy and Geoenvironmental Engineering (1999), both at the Pennsylvania State University. He won the Howard N.Eavenson Award of SME in 2000.

DON W.STEEPLES, is currently Dean A.McGee Distinguished Professor of Applied Geophysics, Department of Geology at the University of Kansas, and president of Great Plains Geophysical, Inc. Previously, he held positions at the Kansas Geological Survey. He holds a Ph.D. in geophysics from Stanford University. Dr. Steeples is involved in the development and application of noninvasive geophysical techniques, specifically, shallow seismic reflection methods applied to environmental and groundwater problems. He has served on several NAS/NRC committees, such as the Committee for Noninvasive Characterization of the Shallow Subsurface for Environmental and Engineering Applications, the Geotechnical Board, and the Committee to Examine the Research Needs of the Advanced Extraction and Process Technology Program. He has published more than 100 articles on the application of geophysical methods and is currently an editorial referee for more than 20 scholarly journals, including *Geophysical Journal International, Geology, Geophysics, Journal of Applied Geophysics, Journal of Contaminant Hydrology, Journal of Environmental and Engineering Geophysics,* and *Journal of Geophysical Research.*

CLINTON L.STRACHAN, P.E., senior geotechnical engineer, Shepherd Miller Incorporated, is a civil engineer with a specialty in geotechnical engineering. Mr. Strachan has been involved with mining and geotechnical engineering projects worldwide. Project experience includes site selection, site exploration, material characterization, design, permitting, construction, and reclamation. Mr. Strachan holds a B.S. in agricultural engineering

(1972) and a M.S. in civil engineering (1979), both from Colorado State University. He is a member of the American Society of Civil Engineers and the Society of Mining Engineers, and chair of the Tailings Dam Committee, U.S. Society on Dams (formerly the U.S. Commission on Large Dams). Mr. Strachan has authored several papers on mine facilities, site investigation, design, and reclamation.

RICHARD J.SWEIGARD, P.E., is chair and professor in the Department of Mining Engineering, University of Kentucky. Prior to his academic positions, he was an engineer for Consol Coal Company and a consulting engineering geologist. Dr. Sweigard's research falls under the category of environmental impacts of mining, including the alleviation of excessive compaction of reconstructed soil, postmining land use, slope stabilization on abandoned mine lands, and disposal of coal combustion by-products. He is a registered engineer in Pennsylvania. His professional activities include membership in the Society of Mining, Metallurgy, and Exploration, and the American Society for Surface Mining and Reclamation; and formerly the American Society of Civil Engineers.

JACK TISDALE is a self-employed coal mine safety specialist, with a specialty in mine safety programs, Mine Safety and Health Administration (MSHA) regulations, and accident analysis. He is retired from MSHA, where he became chief of the Safety Division and program manager of the Accident Investigation (1991–1997) and worked as a safety specialist (1988– 1991). He received the Department of Labor Award for Distinguished Career Service (1994). While working at MSHA, he also directed several major MSHA special projects including the Bleeder and Gob Ventilation Training Course, the Surface Haulage Task Force, and the implementation of the 1992 Mine Ventilation Regulations. Previously, he developed safety and training programs while working for several companies within the coal mining industry, including Eastern Associated Coal Corporation (1977– 1982), Pennsylvania Mines Corporation (1982–1985), and Island Creek Corporation (1986–1988). Mr. Tisdale earned a B.S. degree in mining engineering from the University of Illinois. He served as a commissioned officer with the U.S. Navy Reserve (1955–1957). He was a federal coal mine inspector for the Bureau of Mines in Kentucky, Virginia, and Indiana (1962– 1971) and the sub-district manager and district manager in Pennsylvania and northern West Virginia (1971–1977).

DAVID R.WUNSCH, P.O., is currently the state geologist and director of the New Hampshire Geological Survey. He is also an adjunct professor at

Dartmouth College and the University of New Hampshire. Previously, he was the coordinator of the Coal-Field Hydrology Program at the Kentucky Geological Survey, and an adjunct professor at the University of Kentucky, where he taught courses in applied hydrogeology and low-temperature geochemistry. He holds a Ph.D. in hydrogeology from the University of Kentucky, Lexington. His area of expertise includes mine hydrology and reclamation, geologic hazards, groundwater exploration, and groundwater geochemistry. Dr. Wunsch served as a congressional science fellow, where he advised the U.S. House of Representatives' Subcommittee on Energy and Mineral Resources. He has authored more than 40 technical publications related to coalfield hydrology, and has served on numerous state and federal task forces and committees, including the Kentucky Ground Water Monitoring Committee, and an OSM task force charged with preparing a technical guidance document to aid in the prevention of hydraulic blow-outs from underground coal mines. He is a member of several professional organizations, including the Association of American State Geologists, the American Geophysical Union, the Geological Society of America, Sigma Xi, and the Association of Ground Water Scientists and Engineers. Dr. Wunsch was chosen as Outstanding Kentucky Geologist, 1999, by the Kentucky Chapter of the American Institute of Professional Geologists, and by the John Webster Foster Memorial Lecturer, Illinois State University. Dr. Wunsch is a registered professional geologist in the Commonwealth of Kentucky.

NRC Staff

TAMARA L.DICKINSON, *study director,* is a Senior Program Officer with the National Research Council's Board on Earth Sciences and Resources, responsible for managing the Earth Resources activities of the Board. She has served as program director for the Petrology and Geochemistry Program in the Division of Earth Sciences at the National Science Foundation. She has also served as discipline scientist for the Planetary Materials and Geochemistry Program at NASA Headquarters. As a post-doctoral fellow at the NASA Johnson Space Center, she conducted experiments on the origin and evolution of lunar rocks and highly reduced igneous meteorites. She holds a Ph.D. and an M.S. in geology from the University of New Mexico and a B.A. in geology from the University of Northern Iowa.

KAREN L.IMHOF is a senior project assistant for the Board on Earth Sciences and Resources of the National Research Council. She previously worked on the Board on Agriculture and Natural Resources. Before coming to the Academies, she worked as a staff and administrative assistant in diverse organizations, including the National Wildlife Federation, and the Three Mile Island nuclear facility.

KRISTEN L.KRAPF is a research associate for the Board on Earth Sciences and Resources of the National Research Council. She holds a B.A. and an M.S. in environmental sciences from the University of Virginia. Previously, she was the director of programs at the Renewable Natural Resources Foundation in Bethesda, Maryland.

MONICA R.LIPSCOMB is a research assistant for the Board on Earth Sciences and Resources of the National Research Council. She is completing a master's degree in urban and regional planning at Virginia Polytechnic Institute. Previously, she served as a Peace Corps volunteer in the Ivory Coast and worked as a biologist at the National Cancer Institute. She holds a B.S. in environmental and forest biology from the State University of New York at Syracuse.

KERI H.MOORE is a research associate for the Board on Earth Sciences and Resources at the National Research Council. She holds an M.S. in geology from the Colorado School of Mines and a B.S. in geology with a minor in Russian studies from the College of William and Mary. Previously, she worked as a consulting geologist for mineral exploration companies in Denver and Vancouver.

WINFIELD SWANSON is a self-employed editorial consultant who writes, edits, does research, manages projects, and compiles indexes primarily for the scientific community in Washington, DC. From 1984 until 1995, she served as managing editor of the National Geographic Society's scholarly quarterly, *Research & Exploration.* She has a B.A. in biology from Adelphi University.

Appendix B

Information Provided to the Committee

Speakers at Committee Meetings

Terry Ackman, U.S. Department of Energy, Pittsburgh, Pennsylvania
John Ailes, Division of Environmental Protection, Nitro, West Virginia
David Akers, CQ, Inc., Homer City, Pennsylvania
Dave Altizer, A&L Surveying, London, Kentucky
Barbara Arnold, PrepTech Inc., Apollo, Pennsylvania
Anne S.Barth, Senator Robert C.Byrd's Office, Charleston, West Virginia
Joe Battista, Cofiring Alternatives, Ebensburg, Pennsylvania
Richard Benson, Technos, Inc., Miami, Florida
Hugh E.Bevans, U.S. Geological Survey, Water Resources Division, Charleston, West Virginia
Jason Bostic, West Virginia Coal Association, Charleston
Pat Brady, Mine Safety and Health Administration, Mt. Hope, West Virginia
Gail Montgomery Brion, University of Kentucky, Lexington
Tracy D.Branam, Indiana Geological Survey, Bloomington
Jeffrey D.Brock, Alliance Coal, Lexington, Kentucky

Lech S.Brzezinski, LSB Consulting Services, Nun's Island, Quebec, Canada
Ed Burns, Coastal Coal, Coeburn, Virginia
Roger Calhoun, Office of Surface Mining, Charleston, West Virginia
Matt Cartier, Pittston Coal, Dante, Virginia
Y.Paul Chugh, Southern Illinois University, Carbondale, Illinois
Brad Cole, Pennsylvania Bureau of Deep Mine Safety, Uniontown
Greg Conrad, Interstate Mining Compact Commission, Herndon, Virginia
Bill Cook, Division of Environmental Protection, Logan, West Virginia
Nova "Jodi" Cooper, Big Coal River Watershed Association, Whitesville, West Virginia
David Cowherd, CBC Engineers & Associates, Centerville, Ohio
John Craynon, Office of Surface Mining, Washington, DC
Bob Crowder, R.E.Crowder Consulting, Golden, Colorado
Jeff Daniels, The Ohio State University, Columbus
Pierre DeBlois, Leica Geosystems, Lachine, Quebec, Canada
Greg Demyan, Division of Environmental Protection, Oak Hill, West Virginia
Allen Drake, Phoenix Process Equipment, Louisville, Kentucky
Steven H.Dula, Catenary Coal, Eskdale, West Virginia
Ian Duncan, Division of Mineral Resources, Charlottesville, Virginia
James F.Durant, Alstom Power, Windsor, Connecticut
Rick Eades, Ohio Valley Environmental Coalition, Charleston, West Virginia

Larry Emerson, Arch Coal, Huntington, West Virginia

Erkan "Doc" Esmer, Esmer and Associates, Boomer, West Virginia

Nick Fedorko, West Virginia Geological and Economic Survey, Morgantown

Nathan Petty, West Virginia Rivers Coalition, Elkins

Julia Fox, Marshall University, Huntington, West Virginia

J.Steven Gardner, Engineering Consulting Services, Inc., Lexington, Kentucky

Robert F.Gates, West Virginia Highlands Conservancy, Charleston, West Virginia

Daniel J.Geiger, James River Coal Service, London, Kentucky

Lawrence Gochioco, GX Technology Corporation, Houston, Texas

Richard E.Gray, GAI Consultants, Monroeville, Pennsylvania

Steve Greb, Kentucky Geological Survey, Lexington

Ed Greenwald, Jr., Eavenson, Auchmuty, and Greenwald, Lawrence, Pennsylvania

William E.Griffith, Jr., Division of Environmental Protection, Nitro, West Virginia

Roland Gritto, Lawrence Berkeley National Laboratory, Berkeley, California

Allen Hatheway, University of Missouri-Rolla (professor *emeritus*), Rolla

Michael Hicks, Marshall University, Huntington, West Virginia

Tom Hines, Ohio Division of Mineral Resources Management, Columbus

Anthony lannacchione, National Institute for Occupational Safety and Health, Pittsburgh, Pennsylvania

Gordon Ingram, Arclar Coal Company, Harrisburg, Pennsylvania
William Johnson, D'Appolonia, Monroeville, Pennsylvania
Cynthia M.Kane, U.S. Fish and Wildlife Service, Gloucester, Virginia
Ava C.King, Division of Environmental Protection, Nitro, West Virginia
Jim Kipp, University of Kentucky, Kentucky Water Resources Research Institute, Lexington
O.Eugene Kitts, Summit Engineering, Charleston, West Virginia
Robert Kleinmann, National Energy Technology Laboratory, Pittsburgh, Pennsylvania
Mike Lawless, Mine Safety and Health Administration, Arlington, Virginia
Bruce Leavitt, Consulting Hydrogeologist, Washington, Pennsylvania
Allen Luttrell, Kentucky Department for Surface Mining, Reclamation, and Enforcement, Frankfort
Joseph Main, United Mine Workers of America, Fairfax, Virginia
Doug D.Moore, Addington Resources, Ashland, Kentucky
Dale Morgan, Massachusetts Institute of Technology, Cambridge, Massachusetts
Charles H.Norris, Geo-Hydro, Denver, Colorado
John O'Hare, Arch Coal Company, Huntington, West Virginia
Robin Oder, EXPORTech Company, New Kensington, Pennsylvania
Gary Olhoeft, Colorado School of Mines, Golden, Colorado
Greg A.Olyphant, Indiana University, Bloomington
B.K.Parekh, University of Kentucky, Lexington

Joe Parker, Division of Environmental Protection, Oak Hill, West Virginia

Jim Pierce, Division of Environmental Protection, Oak Hill, West Virginia

William S.Plassio, Department of Environmental Protection, McMurray, Pennsylvania

Dan J.Poteet, EnviroPower, Lexington, Kentucky

Greig J.Robertson, Office of Surface Mining, Mine Map Repository, Pittsburgh, Pennsylvania

Ron Robinson, Department of Mines, Minerals, and Energy, Big Stone Gap, Virginia

L.Rick Ruegsegger, Ohio Department of Transportation, Columbus

Dink Shackleford, Virginia Mining Association, Norton

Mark Skiles, Mining Safety and Health Administration, Arlington, Virginia

Steve Slottee, Eimco Process Equipment Co., Salt Lake City, Utah

William E.Smith, Representative (Kentucky) Harold "Hal" Roger's Office, Washington, DC

Dan Sweeney, U.S. Environmental Protection Agency, Philadelphia, Pennsylvania

Allan Tanner, Radon Consultant, Reston, Virginia

Richard Terry, Sedgman, Beckley, West Virginia

Barry Thacker, Geo/Environmental Associates, Knoxville, Tennessee

Cindy L.Tibbott, U.S. Fish and Wildlife Service, State College, Pennsylvania

Dirk Van Zyl, Mackay School of Mines, Reno, Nevada

Glenn Wattley, Stolar Horizon, Inc., Raton, New Mexico

Bryan Watts, Klohn Crippen, Richmond, British Columbia, Canada
Anthony "Tony" J.Widenman III, Detroit Edison, Detroit, Michigan
Freda Williams, Coal River Mountain Watch, Whitesville, West Virginia
Ed Wojtowicz, Division of Environmental Protection, Oak Hill, West Virginia
Kelvin K.Wu, Mine Safety and Health Administration, Pittsburgh, Pennsylvania
Roe-Hoan Yoon, Virginia Polytechnic and State University, Blacksburg
Michael Young, Pennsylvania Coal Association, Pittsburgh
Dave Zegeer, Kentucky Geologic Survey, Lexington
Robert Zik, Teco Coal, Corbin, Kentucky

Speakers at Public Participation Fora

Linda Adkins, Coal River Mountain Watch, West Virginia
Patty Adkins, Huntington, West Virginia
Connie Agnin, Ohio Valley Environmental Coalition, Huntington, West Virginia
Julie Archer, Pettus West Virginia
Lewis Baker, Huntington, West Virginia
Dianne Bady, Ohio Valley Environmental Coalition, Huntington, West Virginia
Lynn Bentley, Inez, Kentucky
Tom Berel, Pettus, West Virginia
Carolyn Biltney, Inez, Kentucky

Don Blakely, Illinois Department of Natural Resources, St. Louis, Missouri
Deborah Blummins, Inez, Kentucky
Julia Bonds, Coal River Mountain Watch, Rock Creek, West Virginia
Charles Bradford, Clarksburg, West Virginia
Edna Brady, Pettus, West Virginia
Dan Carber, Pettus, West Virginia
Matthew Carther, Pittston Coal, Lebanon, Virginia
Burt Cassidy, Inez, Kentucky
William Chandler, Inez, Kentucky
Doyle Coakley, Citizens' Coal Council, Cowen, West Virginia
James Dickens, Sylvester, West Virginia
Rick Eades, Ohio Valley Environmental Coalition, Charleston, West Virginia
Sarah Ealey, Pettus, West Virginia
Laura Forman, Ohio Valley Environmental Coalition, Huntington, West Virginia
Winnie Fox, Ohio Valley Environmental Coalition, Huntington, West Virginia
Dan Geiger, James River Coal Company, Norton, Virginia
Larry Gibson, Stanley Heir's Foundation, Redhouse, West Virginia
Kenneth Harden, Inez, Kentucky
Adam Holyier, Pettus, West Virginia
Dan Kash, Ohio Valley Environmental Coalition, Ashland, Kentucky

Sister Anne-Marie Listen, Catholic Church, Whitesville, West Virginia

Carl Maronde, U.S. Department of Energy, National Environmental Technology Lab, Pittsburgh, Pennsylvania

Julian Martin, Highlands Conservancy, Charleston, West Virginia

Eugene Massey, Pettus, West Virginia

Janice Maynard, Inez, Kentucky

Nancy McBay, Pettus, West Virginia

John McCormick, Citizens' Coal Council, Washington, DC

Nina McCoy, Kentuckians for the Commonwealth, Inez, Kentucky

John Moore, Pilgrim, Kentucky

Janice Nease, Coal River Mountain Watch, Whitesville, West Virginia

Steve O'Brien, Pettus, West Virginia

Bill Plassio, Pennsylvania Department of Environmental Protection, McMurray, Pennsylvania

Greg Priest, Inez, Kentucky

Melena Preece, Kentuckians for the Commonwealth, Inez, Kentucky

Barney Reilly, Dickenson County Citizens Committee, Norton, Virginia

Vina Reilly, Dickenson County Citizens Committee, Norton, Virginia

Ronald Robinson, Virginia Department of Mines, Minerals and Energy, Big Stone Gap, Virginia

Dink Shackleford, Virginia Mining Association, Norton, Virginia

Jo Skinnex, Student Association for the Environment [SAFE], Ashland, Kentucky

Vivian Stockman, Ohio Valley Environmental Coalition, Huntingdon, West Virginia

James Spencer, Stickney, West Virginia

Carol Tiller, Inez, Kentucky

Patty Wallace, Kentuckians for the Commonwealth, Louisa, Kentucky

Freda Williams, Ohio Valley Environmental Coalition, Leevale, West Virginia

Mike Young, Pennsylvania Coal Association, Harrisburg, Pennsylvania

Individuals Who Provided Technical Documents to the Committee

Terry Ackman, U.S. Department of Energy, Pittsburgh, Pennsylvania

John Ailes, Division of Environmental Protection, Nitro, West Virginia

David Akers, CQ, Inc., Homer City, Pennsylvania

Dave Altizer, A&L Surveying, London, Kentucky

Barbara Arnold, PrepTech Inc., Apollo, Pennsylvania

Anne S. Earth, Senator Robert C. Byrd's Office, Charleston, West Virginia

Joe Battista, Cofiring Alternatives, Ebensburg, Pennsylvania

Richard Benson, Technos, Inc., Miami, Florida

Hugh E.Bevans, U.S. Geological Survey, Water Resources Division, Charleston, West Virginia

Julia Bonds, Rock Creek, West Virginia

Jason Bostic, West Virginia Coal Association, Charleston

Pat Brady, Mine Safety and Health Administration, Mt. Hope, West Virginia

Gail Montgomery Brion, University of Kentucky, Lexington
Tracy D.Branam, Indiana Geological Survey, Bloomington
Gary Brill, Geo/Environmental Associates, Knoxville, Tennessee
Jeffrey Brock, Alliance Coal, Lexington, Kentucky
Lech Brzezinski, LSB Consulting Services, Nun's Island, Quebec, Canada
Moss Burgess, Wilkinson, West Virginia
Ed Burns, Coastal Coal, Coeburn, Virginia
Robert Byrd, U.S. Senator, West Virginia
Roger Calhoun, Office of Surface Mining, Charleston, West Virginia
David Carris, J.T.Boyd, Pittsburgh, Pennsylvania
Matt Cartier, Pittston Coal, Dante, Virginia
Y.Paul Chugh, Southern Illinois University, Carbondale, Illinois
Doyle Coakley, Citizens Coal Council, Cohen, West Virginia
Jim Cobb, Kentucky Geological Survey, Lexington
Brad Cole, Pennsylvania Bureau of Deep Mine Safety, Uniontown
Greg Conrad, Interstate Mining Compact Commission, Herndon, Virginia
Bill Cook, Division of Environmental Protection, Logan, West Virginia
David Cooper, Sierra Club, Lexington, Kentucky
Nova "Jodi" Cooper, Big Coal River Watershed Association, Whitesville, West Virginia
David Cowherd, CBC Engineers & Associates, Centerville, Ohio

John Craynon, Office of Surface Mining, Washington, DC
Bob Crowder, R.E.Crowder Consulting, Golden, Colorado
Jeff Daniels, The Ohio State University, Columbus
Pierre DeBlois, Leica Geosystems, Lachine, Quebec, Canada
Greg Demyan, Division of Environmental Protection, Oak Hill, West Virginia
Allen Drake, Phoenix Process Equipment, Louisville, Kentucky
Gary Drake, Phoenix Process Equipment, Louisville, Kentucky
Steven H.Dula, Catenary Coal, Eskdale, West Virginia
Ian Duncan, Division of Mineral Resources, Charlottesville, Virginia
James F.Durant, Alstom Power, Windsor, Connecticut
Rick Eades, Ohio Valley Environmental Coalition, Charleston, West Virginia
Larry Emerson, Arch Coal, Huntington, West Virginia
Erkan "Doc" Esmer, Esmer and Associates, Boomer, West Virginia
Nick Fedorko, West Virginia Geological and Economic Survey, Morgantown
Nathan Fetty, West Virginia Rivers Coalition, Elkins
Tony Fonseca, Consol (retired), Pittsburgh, Pennsylvania
Julia Fox, Marshall University, Huntington, West Virginia
Steve Gardner, Engineering Consulting Services, Inc., Lexington, Kentucky
Robert F.Gates, West Virginia Highlands Conservancy, Charleston, West Virginia
Daniel J.Geiger, James River Coal Service, London, Kentucky

Charles Gillian, Alliance Consulting Company, Beaver, West Virginia

Lawrence Gochioco, GX Technology Corporation, Houston, Texas

Richard E.Gray, GAI Consultants, Monroeville, Pennsylvania

Steve Greb, Kentucky Geological Survey, Lexington

Ed Greenwald, Jr., Eavenson, Auchmuty, and Greenwald, Lawrence, Pennsylvania

William E.Griffith, Jr., Division of Environmental Protection, Nitro, West Virginia

Roland Gritto, Lawrence Berkeley National Laboratory, Berkeley, California

Allen Hatheway, University of Missouri-Rolla (professor emeritus), Rolla

Michael Hicks, Marshall University, Huntington, West Virginia

Barry Hillman, Condor Earth, Sonora, California

Tom Hines, Ohio Division of Mineral Resources Management, Columbus

Rick Honaker, University of Kentucky, Lexington

Evan Hughes, EPRI, Upgraded Coal Interest Group, Palo Alto, California

Marshall Hunt, Consol Inc., Pittsburgh, Pennsylvania

Anthony Iannacchione, National Institute for Occupational Safety and Health, Pittsburgh, Pennsylvania

Gordon Ingram, Arclar Coal Company, Harrisburg, Pennsylvania

John B.Ivey, Amuedo and Ivey, Inc., Arvada, Colorado

William Johnson, D'Appolonia, Monroeville, Pennsylvania

William Kalb, Tra-Det, Wheeling, West Virginia

Cynthia M.Kane, U.S. Fish and Wildlife Service, Gloucester, Virginia

Ava C.King, Division of Environmental Protection, Nitro, West Virginia

Jim Kipp, University of Kentucky, Kentucky Water Resources Research Institute, Lexington,

Eugene Kitts, Summit Engineering, Charleston, West Virginia

Robert Kleinmann, National Energy Technology Laboratory, Pittsburgh, Pennsylvania

Mark Klima, Pennsylvania State University, State College

Kewal Kohli, Office of Surface Mining, Pittsburgh, Pennsylvania

John Kramer, Condor Earth, Sonora, California

Mike Lawless, Mine Safety and Health Administration, Arlington, Virginia

Bruce Leavitt, Consulting Hydrogeologist, Washington, Pennsylvania

Thomas Lefchik, Federal Highway Administration, Columbus, Ohio

Henry Liu, University of Missouri, Columbia

Allen Luttrell, Kentucky Department for Surface Mining, Reclamation, and Enforcement, Frankfort

Joseph Main, United Mine Workers of America, Fairfax, Virginia

Carl Maronde, Department of Energy, Pittsburgh, Pennsylvania

Mikhail Maryamchik, Babcock and Wilcox, Barberton, Ohio

Nina McCoy, Inez, Kentucky

Jim McElfish, Environmental Law Institute, Washington, D.C.

John Mead, Southern Illinois University, Carbondale

Bruce Miller, Pennsylvania State University, State College

Keith Mohn, Lone Mountain Processing, Inc., Pennington Gap, Virginia

Doug Moore, Addington Resources, Ashland, Kentucky

Dale Morgan, Massachusetts Institute of Technology, Cambridge, Massachusetts

Joel Morrison, Pennsylvania State University, State College

Janice Nease, Coal River Mountain Watch, Whitesville, West Virginia

Charles H.Norris, Geo-Hydro, Denver, Colorado

John O'Hare, Arch Coal Company, Huntington, West Virginia

Robin Oder, EXPORTech Company, New Kensington, Pennsylvania

Gary Olhoeft, Colorado School of Mines, Golden, Colorado

Greg Olyphant, Indiana University, Bloomington

B.K.Parekh, University of Kentucky, Lexington

Joe Parker, Division of Environmental Protection, Oak Hill, West Virginia

Brenda Pierce, U.S. Geological Survey, Reston, Virginia

Jim Pierce, Division of Environmental Protection, Oak Hill, West Virginia

William Plassio, Department of Environmental Protection, McMurray, Pennsylvania

Dan Poteet, EnviroPower, Lexington, Kentucky

Cliff Raleigh, CQ Inc., Homer City, Pennsylvania

Greig Robertson, Office of Surface Mining, Mine Map Repository, Pittsburgh, Pennsylvania

Ron Robinson, Department of Mines, Minerals, and Energy, Big Stone Gap, Virginia

L.Rick Ruegsegger, Ohio Department of Transportation, Columbus

Andy Schissler, Stolar Horizon, Raton, New Mexico

Dink Shackleford, Virginia Mining Association, Norton

Mark Skiles, Mining Safety and Health Administration, Arlington, Virginia

Steve Slottee, Eimco Process Equipment Co., Salt Lake City, Utah

Will Smith, Legislative Director, Harold Rogers Office, U.S. House of Representatives, Kentucky

Dean Spindler, Illinois Geological Survey, Springfield, Illinois

Dan Sweeney, U.S. Environmental Protection Agency, Philadelphia, Pennsylvania

Darrell Taulbee, Center for Applied Energy Research, University of Kentucky, Lexington

Allan Tanner, Radon Consultant, Reston, Virginia

Richard Terry, Sedgman, Beckley, West Virginia

Barry Thacker, Geo/Environmental Associates, Knoxville, Tennessee

Cindy L.Tibbott, U.S. Fish and Wildlife Service, State College, Pennsylvania

Larry Trainor, Office of Surface Mining, Washington, D.C.

Dirk Van Zyl, Mackay School of Mines, Reno, Nevada

Jeff Vansant, Division of Waste Management, Commonwealth of Kentucky

Steve Vick, Consultant, Bailey, Colorado

Glenn Wattley, Stolar Horizon, Inc., Raton, New Mexico

Bryan Watts, Klohn Crippen, Richmond, British Columbia, Canada
Bruce Watzman, National Mining Association, Washington, D.C.
Anthony "Tony" J.Widenman III, Detroit Edison, Detroit, Michigan
Freda Williams, Coal River Mountain Watch, Whitesville, West Virginia
Ed Wojtowicz, Division of Environmental Protection, Oak Hill, West Virginia
Mike Wooldridge, Alliance Consulting Company, Beaver, West Virginia
Kelvin K.Wu, Mine Safety and Health Administration, Pittsburgh, Pennsylvania
Roe-HoanYoon, Virginia Polytechnic and State University, Blacksburg
Michael Young, Pennsylvania Coal Association, Pittsburgh
Steve Young, Pipeline Systems Incorporated, Walnut Creek, California
Dave Zegeer, Kentucky Geologic Survey, Lexington
Robert Zik, Teco Coal, Corbin, Kentucky

Appendix C

Glossary

ABANDONED IMPOUNDMENT	An impoundment that is not in operation and is closed. It has been filled to capacity and reclaimed.
ACTIVE IMPOUNDMENT	An impoundment that is in operation and receiving slurry.
AQUITARD	A low permeability geologic horizon that restricts the migration of water under ordinary hydraulic gradients.
ASH	The inorganic residue after burning, esp. of coal. The term is also used informally to refer to the non-coal mineral matter associated with coal that will become ash after combustion at the power plant. Coal with more ash has less heating value.
AUGER	Mining equipment used to extract coal by boring horizontal holes into the coal seam. Auger mining is conducted from the outcrop most typically in conjunction with contour surface mining.
BARRIER PILLAR	Block of coal left in place in an underground mine for safety and stability purposes and to prevent water influx.
BASIN	*(as used in this report)* Area of existing contours within an impoundment that excludes the embankment; the area covered by slurry and water; both the floor and the walls of the pool behind the embankment.
BEACH	*(as used in this report)* The subaerial accumulation of the coarser fraction deposited from slurry.
BED SEPARATION	The parting of strata along upper and/or lower contacts due to subsidence.
BLACK WATER	Water mixed with fine coal refuse.
BLIND FLUSHING	Injection into an underground mine whose access has been obstructed.
BLOW-IN	Opposite of blow-out; catastrophic failure resulting from water pressure building up on the surface of an outcrop coal barrier; the material "blows in" to the underground workings from the surface.

BLOW-OUT	Opposite of blow-in; catastrophic failure resulting from build-up of underground water pressure against an outcrop coal barrier. Material "blows out" from underground workings to the surface.
BONY COAL	Coal high in mineral content, usually clay shale particles.
BREAK-THROUGH	Catastrophic failure/opening of mine working along a fracture, joint, bedding plane, or other zone of weakness.
BTU (BRITISH THERMAL UNIT)	Heating value for a unit weight of coal. The heat needed to raise one pound of water one degree Fahrenheit.
BULK-HEAD	A seal constructed to prevent water from entering or exiting an area of the mine. Barrier design and construction is based upon the amount of water pressure (water head, expressed in feet) that will be exerted against the bulkhead.
CLOSED-LOOP SURVEY	Method of establishing the accuracy of a mine survey by conducting a loop traverse to the point of beginning. The accuracy is measured in terms of a ratio of feet of error to feet of traverse, e.g., 1:5000 indicates an error of 1 foot over a 5000-foot traverse.
COMPRES-SIVE STRESS	Stress that pushes together material on opposite sides of a real or imaginary plane.
CON-TROLLED PLACE-MENT	*(as used in this report)* Injection of coal waste slurry into an underground mine that has been mapped and inspected, whose volume is known, and in which bulkheads may be built to control the direction and extent of flow.
DAM	An artificial barrier or wall constructed across a watercourse to confine flowing water for a variety of purposes, such as creating a pond or lake for storage of water; creating a hydraulic head that can used to generate power; controlling floods; or retention of debris.
DEPTH-TO-MINED-HEIGHT RATIO	Ratio of the depth of the coal seam to the mined-height, or "effective thickness," of the mined seam.
DOWN-STREAM CON-STRUCTION EMBANK-MENT	A method of staged embankment construction where the embankment centerline is moved downstream with subsequent embankment raises.
EMBANK-MENT	*(as used in this report)* A linear engineered structure extending above the natural ground surface that retains fine coal slurry waste material; it is built with earthen materials or coarse coal refuse (waste material).
EXTEN-SOMETER	An instrument for measuring changes caused by stress in a linear dimension of a body.
FACTOR OF SAFETY	A quantitative measure of the ratio of available strength to applied force. For slope stability, factor of safety is the ratio of forces resisting slope movement to forces causing slope movement.

FINES	An informal term referring to fine particles, either product or waste, resulting from processing and preparation; usually less than 100 mesh (150 micrometers) and greater than 325 mesh (45 micrometers).
FOOTPRINT OF THE EMBANKMENT	The area of natural ground to be covered by the embankment.
FREEBOARD	The difference in elevation between the embankment crest or spillway invert (bottom) and the water pool elevation in an impoundment.
FRENCH DRAIN	Small underground channel filled with permeable materials used to convey water passively.
GRANDFATHERED /PRE-LAW IMPOUNDMENT	An impoundment that has not been in operation since promulgation of the 1977 Surface Mining Control and Reclamation Act (SMCRA) regulation. These impoundments are reclaimed under the Abandoned Mine Lands Program.
HYDRAULIC HEAD	The height of the free surface of a body of water above a given subsurface point; or, pressure against the dam from the weight of the slurry *(as used in this report).*
IMPOUNDMENT	*(as used in this report)* The entire structure used for coal slurry waste disposal, including the embankment, basin, beach, pool, and slurry.
INACTIVE IMPOUNDMENT	An impoundment that is not in operation or receiving slurry. Inactive impoundments may receive slurry in the future, becoming active again, and therefore have not been closed permanently.
JOINT	A fracture or parting in rock along which no displacement has occurred.
LIQUEFACTION	The transformation of a solid material, such as loosely packed sediment or cohesionless soil, into a fluid mass due to increased pore pressure and reduced effective stress.
LOADING	Mass and other vertical stresses applied to structure.
MOISTURE CONTENT	*(as used in this report)* The percentage of water in a waste slurry. Calculated as the weight of water divided by the weight of the dried solids multiplied by 100.
MONITORING	Observing, regulating, and evaluating a system to ensure that it is operating properly.
NORTH AMERICAN DATUM 27 (NAD27)	Mapping coordinate standards established in 1927, measured in feet.
NORTH AMERICAN DATUM 83 (NAD83)	Mapping coordinate standards established in 1983, measured in meters.
OUTCROP	The intersection between a geologic formation (e.g., coal seam) and the Earth's surface.

OUTCROP BARRIER (Boundary/ Perimeter Pillar)	Distance between the coal outcrop and the furthest extent of underground mine workings in the direction of the outcrop.
PACKERS	Plugs that are used to isolate fluid under pressure in a specific segment of pipe in a hole.
PARTINGS	Thin sedimentary layers that follow a surface of separation between thicker units of rock.
PERME-ABILITY	The capacity of a porous medium to permit flow of a given fluid.
PHREATIC SURFACE	The groundwater interface or a zone of saturation where the water pressure is equal to atmospheric pressure.
PIPING	Seepage through embankments, which can lead to failure by internal erosion.
POOL	Area of free-standing water separated from the slurry discharge point by the beach; it may contain a low percentage of fines and ultra-fines, and suspended and unconsolidated solids.
PROBA-BLE MAX-IMUM PRECIPI-TATION	The theoretical, greatest depth of precipitation for given duration that is possible over a particular drainage basin.
P-WAVE	Compressional seismic waves; sound waves; they travel faster than other seismic waves.
QUAD-RANGLE MAP, 7 1/2-MINUTE GEOLOGI-CAL	Topographic and geologic maps produced by the U.S. Geological Survey. The maps depict an area of 7 1/2 minutes of latitude and longitude.
RAISE	*(as used in this report)* The construction of an embankment in a staged manner so that the embankment crest is higher in elevation than the previous stage.
REFUSE (OR GOB) PILE	Area where coarse waste material (larger than 28 mesh, 800 micrometers) is disposed.
RUN-OF-MINE	Raw mined material, unaltered from what is transported out of the mine.
SIDE SLOPE	Natural embankment.
SLIMES	Material in the waste stream smaller than 325 mesh (45 micrometers) and composed mainly of clay or clay-like particles; have high moisture content or the ability to retain water.
SLURRY	A mixture of water and solids (less than 28 mesh, 800 micrometer, particle size) prepared for handling as a liquid for processing and disposal.
SOLIDS CONTENT	*(as used in this report)* The percentage of solids in a waste slurry.
SPAD	A flat spike, hammered into a wooden plug, anchored in a hole drilled into the mine ceiling from which is threaded a plumbline. The spad is an

	underground survey station similar to the use of stakes in marking survey points on the surface. A brass tag permanently attached to the mine roof.
STACKING	Disposing of dewatered coal waste in "stacks" or layers (from 4–6 inches to 1–3 feet in thickness) piled on top of each other without an embankment.
STARTER DAM	The initial embankment constructed as the first stage of a staged embankment construction system.
STRATA	Multiple layers of sedimentary rock.
SURFACE WAVES	Seismic waves that travel only near the surface of the earth; they travel about 90 percent as fast as *S*-waves.
S-WAVE	A wave reflecting shear motion, with oscillation perpendicular to the direction of propogation.
TAILINGS	Fine particle waste streams from either coal preparation or other mineral processing plants.
TAILINGS DAM	*(as used in this report)* A structure constructed to contain fine particle waste streams from other mineral processing plants.
TENSILE STRESS	A normal stress causing separation across the plane on which it acts.
TOE DRAIN	A zone of permeable materials constructed at the downstream toe of an embankment to collect and convey water from the downstream region of the embankment.
ULTRA-FINES	Fine particles, either product or waste, resulting from the processing or preparation of any mineral; particles are smaller than 325 mesh (45 micrometers).
UP-STREAM CON-STRUCTION DAM	A method of staged embankment construction where the embankment centerline is moved upstream with subsequent embankment raises.

Appendix D

Acronyms and Abbreviations

CFR	Code of Federal Regulations
DOE	Department of Energy
EPA	Environmental Protection Agency
ICOLD	International Commission on Large Dams
MSHA	Mining Safety and Health Administration
OSM	Office of Surface Mining
SMCRA	Surface Mining Control and Reclamation Act of 1977
SO_2	Sulfur dioxide
USC	United States Code
USCOLD	U.S. Commission on Large Dams

Appendix E

Geophysical Techniques

ELECTRICAL AND ELECTROMAGNETIC METHODS

The induced polarization (IP) method includes analysis of the Earth's delayed response to an induced current; induced polarization is related to the resistivity method. The "P" in IP can be thought of as "persistence" or the amount of time the Earth stays disturbed electrically after the removal of the electrical disturbing function. The discharge rate of a volume of the Earth is similar to that of a capacitor. The induced voltage's decay rate is dependent on the ion mobility in the charged volume. For example, the ions in clays are very mobile. Measurements can be made in the frequency domain, where the phase delays of various frequencies are analyzed, or in the time domain where voltage is measured as a function of time. Highly accurate clocks can be synchronized each day to determine the amount of delay for each frequency reaching the voltage electrodes, or the receiver and transmitter can be connected to a single clock. The typical induced polarization frequency range is between 0.05 hertz and 1 kilohertz. Induced polarization surveys have been used in some cases for groundwater exploration and the method is frequently used in sulfide mineral exploration. Some recommendations for potentially fruitful areas for induced polarization research are given in Ward et al. (1995).

The spontaneous potential method employs natural voltages resulting from electrochemical activity in the Earth. The voltages usually average to zero over distances a few times larger than the spatial extent of any anomalies, and they rarely exceed 100 millivolts. Spontaneous potentials can be generated by fluid, ions, or heat moving in the Earth. The source current or configuration remains unchanged over the period of measurement. Because this is a passive technique the signals are vulnerable to "noise" from powerlines, pipelines, electrical storms, and other environmental sources. Sometimes the noise level may preclude the repeatability of the measurements, which is one of the problems with spontaneous potential methods. Spontaneous potential measurements have been used with some success in geothermal exploration and to monitor subsurface water movement (i.e., observing a moving conductor in a magnetic field). In the geothermal case, chemical reactions induced by mineralized waters may add to any voltage caused by movement of water. Another possible use of spontaneous potential surveys is mapping the concentration gradients of chemically active leachate. A spontaneous potential survey might be sensitive to water in mine works that are reacting with their surroundings. Spontaneous potential data may be interpreted from contour maps of voltages or by more quantitative model calculations employing geometrical shapes similar to those used in magnetic and gravity studies. The fundamentals of near-surface spontaneous potential applications can be found in Corwin (1990).

Active electromagnetic surveying employs a primary field induced by an electrical current passing through a coil, which induces three-dimensional currents through underground conductors according to the physical laws of electromagnetic induction. This underground current induces a secondary electromagnetic field which then distorts the primary field; the resulting final field is sensed by a receiving coil. The field detected by the receiving coil varies in intensity, phase, and direction from the primary field, which reveals information about subsurface electrical conductivity. Electromagnetic methods have an advantage over DC resistivity because they do not require inserting electrodes into the ground. Low-flying, small (maximum dimension 3 to 6 feet) unmanned aircraft have been employed to conduct some surveys to provide access to polluted, dangerous, or inaccessible areas. Such airborne surveys have disadvantages, including small separation between the source and receiver coils and a higher noise level caused by the relatively high velocity movement of the coils through the Earth's magnetic field. The practical background and the theoretical basis for electromagnetic methods are presented in McNeill (1990).

In contrast to active electromagnetic surveying discussed above, passive electromagnetic surveying employs Earth's natural electromagnetic fields to provide the variations in the electric field. Electric fields generated by distant lightning flashes are the source used in the audio-frequency magnetic field technique. The passive very-low radio frequency method relies upon the 15 to 25 kilohertz electric field from distant, powerful radio transmitters used to communicate with submarines. These passive techniques may be useful for regional studies, but they do not have the resolution to find underground mine works.

IN-SEAM SEISMIC TECHNIQUES

Data Acquisition

In-seam seismic surveys are typically performed in panels surrounding blocks of coal prior to long-wall mining operations. Seismic-wave transmission surveys are set up to test the transmissivity of the coal seam by deploying seismic sources along one face of a coal panel and placing geophones along the opposing face. If disturbances are inferred from the transmission experiment, a seismic reflection survey may be used to estimate their locations. Small explosive charges deployed in horizontal holes drilled about 3 feet into the face of the coal seam are used to generate elastic waves within the coal seam (Dresen and Reuter, 1994). Geophones are routinely wedged into horizontal holes similar to the source holes. The geophones are sensitive to motion in the plane of the coal seam, and are designed to record channel waves of Love type (see below). Although coal has a relatively high rate of seismic-wave attenuation (Q factors range from 20 to 50), divergence of in-seam waves is two-dimensional; thus propagation distances as long as a 11/4 mile have been reported (Greenhalgh et al., 1986).

Channel Waves

A coal seam is a low-velocity channel for elastic waves. If a seismic source is triggered in the middle of the coal seam, elastic waves propagate in all directions throughout the coal. Wave motion encountering the coal-rock interface along the top and bottom of the coal seam is constructively reflected back into the seam at various angles and at different phase velocities. This constructive-interference system is a channel wave that propagates within the coal seam without radiating significant wave energy into the surrounding bedrock. The two types of seismic waves, commonly interpreted as part of in-seam seismic surveys, are Rayleigh waves comprised of body waves of the *P* and *SV* type and Love waves comprised of *SH* waves only. Seismic-wave phases created at various angles of reflection at the coal-bedrock interfaces cause dispersion of the channel waves, which means that the channel waves propagate at frequency-dependent speeds. Hence, at longer travel distances the wave phases get separated and are recorded as a time- elongated arrival in the seismogram.

Data Processing

Analysis of in-seam wave dispersion can help determine whether a propagation path of a transmitted wave is disturbed by geologic or old mine features. The seismogram is transformed into a velocity-frequency diagram, where the dispersed channel waves appear as curved-amplitude plots. Because the seismic velocities and densities of both coal and bedrock can be measured easily in coal mines, accurate theoretical dispersion curves can be calculated for the rock-coal-rock geology and plotted for comparison with the seismic velocity-frequency data (Raeder et al., 1985). If the theoretical dispersion curve and field data velocity-frequency plots match closely, an undisturbed propagation path is indicated. However, if the channel waves are reflected from an obstruction back to the panel containing the seismic source, the match may be poor. If a disturbance is inferred, a seismic reflection survey conducted with both the seismic source and geophones at the same seam face may help to estimate the location of the disturbance. Because the seismic-wave velocity in the coal is known, the two-way travel time of any waves reflected from the disturbance back to the geophones can be transformed into distance to indicate the location of the reflector.

NUCLEAR MAGNETIC RESONANCE

Nuclear magnetic resonance measurements were initially performed by physicists investigating molecular-scale phenomena. A radio-frequency pulse excites nuclei to a higher energy state and their return to the original state is monitored, modeled as a sum of exponential decays, and recorded as two relaxation-time constants, T_1 associated with the longitudinal component of the magnetization, and T_2 with the transverse component. Nuclear magnetic resonance can be used to study any nuclei that have an intrinsic magnetic moment, such as hydrogen or carbon-13. (See McMurray's 1984 review of nuclear magnetic resonance theory.)

Surface geophysical nuclear magnetic resonance was pioneered by a Russian team (Semenov et al., 1988) who developed the "hydroscope" consisting of a transmitter and receiver in which antennas approach 330 feet in diameter (a lower limit on horizontal spatial resolution). The total volume of water present as a function of depth is proportional to the number of hydrogen nuclei in the sample which is proportional to the amplitude of the initial magnetization. When the transmitter is turned off, the resulting relaxation time contains information about the grain size of the water-saturated rock. If the rock or soil contains water, the relaxation time is a function of two processes: relaxation in the bulk fluid and relaxation on the solid pore surfaces. Surface relaxation is the faster of the two processes, dominating the response, and leading to a relationship between relaxation time and the ratio of the pore's surface area to volume, which is related to grain size. An empirical correlation between rock type and decay time was used by Shirov et al. (1991) to estimate grain size. Minimization of the misfit between the calculated and measured responses is via inversion is used to extract both the total amplitude and the relaxation-time constants.

Paramagnetic species (such as Fe^{3+}) can cause dramatic changes in T_2 so that the direct nuclear magnetic resonance link to the ratio of the surface area to volume breaks down. These effects, which make it much more difficult to obtain estimates of permeability, were examined by Bryar et al. (2000) and Knight et al. (1999). For example, two sands whose pore size and distribution and grain size are identical could appear to have different permeabilities if one had a high Fe^{3+} content and the other did not. The variation in the content of Fe^{3+} and other paramagnetic species could complicate or negate permeability estimates based on NMR data for near-surface applications. T_2 would be affected, but to a different extent, depending on the specific location of the Fe^{3+} (i. e., in pore water, adsorbed to the solid phase, or in a solid mineral grain).

BOREHOLE GEOPHYSICS

The vast majority of surface geophysical techniques can be modified for application in borehole environments. This includes resistivity, electromagnetic, gravity, radar, radiometric, and seismic methods (e.g., Daniels and Keys, 1990; Howard, 1990).

In some cases, borehole geophysical measurements are made to help tie the borehole samples to surface geophysical data. In other cases, logs are used to help interpret the samples themselves. For example, in the petroleum industry, to calculate the hydrocarbon concentration, the- resistivity log is used to infer the percentage of saturation with hydrocarbons, where the salinity of the formation water is known.

Sometimes a particular log will be diagnostic in a particular environment, and other times the geology will defy rational analysis by even the most sophisticated suites of logs. On balance, however, it is remarkable how much geologic information can be derived from simple suites of logs, given the gross physical assumptions that are made in logging.

Properties that can often be directly or indirectly determined from borehole geophysics include, but are not limited to, the following:

1. Lithology
2. Bed thickness
3. Porosity
4. Fluid type and amount
5. Fluid flow vector
6. Permeability
7. Trace element chemistry
8. Fracture orientation
9. Rock strength

Discussed below are logging techniques to infer these properties. Problems in geophysical logging of oil wells include the presence of mud cake and of formation disturbance by drilling fluids. The presence of these materials disturbs the geophysical measurements.

Density Logs

The density log employs a cesium source of gamma rays shot into the formation at 45° away from the hole axis. The receiver is a scintillation counter collimated at 45° to the hole also but at a right angle to the source direction. The method uses Compton scattering and assumes a direct relationship between electron density and bulk density. This is actually a surprisingly accurate assumption because the ratio is very close to 2:1 for mass compared to number of electrons. The level of gamma radiation caused by the scattering is proportional to the total number of electrons in the formation near the sonde.

Neutron Logs

Nonradioactive elements emit gamma radiation if they are sprayed by a stream of neutrons. The neutron source for logging is americium or beryllium, which produces a constant population of high-energy neutrons. Hydrogen selectively absorb these neutrons, and as the neutrons are absorbed, gamma rays are emitted. The more hydrogen in the rock, the faster neutrons are absorbed and the more gamma rays are emitted. Water or oil absorbs neutrons; therefore, porosity is generally proportional to hydrogen content as indicated by the neutron log. The neutron log is a porosity determination tool.

Resistivity Logs

The resistivity log is analogous to electrical resistivity measurements made at the Earth's surface. It measures the electrical resistance of material between electrodes placed on the sonde in the borehole. This log is especially sensitive to the electrical properties of fluids contained in underground formations. The resistivity log cannot generally be made through casing, although research is now being done to develop this capability. There are several types of resistivity logs including direct-current resistivity logs and electromagnetic induction logs.

Gamma Logs

Gamma logs measure natural gamma radiation, and are particularly useful for finding shales that have a high gamma output because clay collects radionuclides. They can be used in cased holes, so they are also run on casing collar logs to tie the exact location of casing to geologic rock units. This is essential if we are to perforate the casing at exactly the right spot to test the oil or gas (or fresh water) zones. Newer techniques include gamma-ray spectral logging to look at clays, based on ratios of gamma rays of known energy from uranium, thorium, and potassium.

Spontaneous Potential Logs

The spontaneous potential logs measure voltage between formations by attaching one voltmeter electrode on the logging sonde and the other at the Earth's surface (see above). It does not work if the drill is using salt-based mud.

Sonic Logs

Sonic logs are essentially a borehole seismic refraction survey. Sonic logs use a 20-kilohertz transducer and two sensors. The method makes use of the Wyllie equation which assumes that transit time is a function of the mineralogy, the percentage of pore space, and the *P*-wave velocity of the fluid within the pores. The equation works surprisingly well, but it is sometimes treacherous to extrapolate the velocity measured by sonic tools to velocity measured by seismic waves that have wavelengths 1,000 times as long.

The analyst measures the transit time in microseconds per foot to determine rock type and porosity.

Temperature Logs

The temperature log is useful for measuring temperature gradients, lithology changes, and water flow in the vertical direction. The logging can be done either from top down or from bottom up, and it is common to log in both directions. However, for the greatest precision, logging from the top down is preferred because the water has not been disturbed by the passing of the tool.

Caliper Logs

The caliper log is used to measure borehole diameter as a function of depth. It shows the boundaries between soft shales and hard limestones very clearly and with better depth, precision than other logging tools. It is useful to find evaporites and washouts of shale.

Casing Collar Logs

The casing collar log is used to count joints of casing to know exactly how far down the hole specific geologic layers are. The casing collar log is used in conjunction with the natural gamma log to provide locations for casing perforations or for hydrologic measurements such as packer tests and drill stem tests. (Packers are plugs used to isolate fluid under pressure in a specific segment of pipe in a hole.)

Dip-Meter Logs

The dip-meter log is obtained from three resistivity tools placed at different azimuths around the sonde. This log of measures local dip of geologic layers within a borehole.

Cement Bond Logs

The cement bond log is obtained with a sonic tool to determine how good the cement seal is on the outside of the casing. The highest amplitude is related to poor or nonexistent bonds between the casing and the cement. Low amplitude indicates good bonding, which allows the sound energy to penetrate away from the casing. A good bond ensures against leaks of water or pollutants from one rock layer to another.

Borehole Acoustic Televiewer Logs

The borehole acoustic televiewer log is used to develop acoustical analog images of the rock face around the borehole. This is particularly useful for determining fracture patterns and directions. The fracture directions are needed when designing horizontal drilling programs where the bit must cross fractures to drain oil reservoirs efficiently. It can be used in drilling mud. Also, this tool can be used in conjunction with hydraulic fracturing to determine in situ stress orientation of the principal tectonic stresses. The hole is first surveyed with a televiewer. Then packers are set at the top and bottom of the interval of interest and the formation is hydraulically fractured by injecting water into the interval between the packers. Finally, the packers are removed and the hole is inspected again by televiewer to determine the direction the fracture orientations. Fracturing occurs perpendicular to the direction of least principal stress.

Borehole Television Camera

The borehole television camera is used to develop a visual image of the borehole walls. It is used for many of the same things as the acoustic televiewer, but the water in the hole must be relatively clear for it to work. When it works, it can provide a more detailed image of fractures and even of lithology than the acoustic televiewer.

Drill Penetration Logs

The rate of drill penetration into the ground is measured with a drill penetration log. It is used in conjunction with logs of mud-pump pressure, rate of spin of the bit, and weight on the bit. In areas where previous experience is available, these logs can be extremely useful in knowing where the bit is geologically. This log is a measurement-while-drilling log. Other measurement-while-drilling tools, such as resistivity tools, are also available to help prevent blow-outs. They detect highly electrically resistive conditions (e.g., overpressured gas) ahead of the bit.